Health and Safety Testing in Construction

Edition 6

CSC 01/06

Published by CITB-ConstructionSkills
Bircham Newton, King's Lynn
Norfolk PE31 6RH

First Published 2000

Revised 2001
Revised 2002
Revised 2003
Revised 2004
Revised 2005

© Construction Industry Training Board 2000

ISBN 1 85751 106 9

CITB-ConstructionSkills has made every effort to ensure that the information contained in this publication is accurate. It should be used as guidance material and not as a replacement for current regulations or existing standards.

All rights reserved. No part of this publication may be reproduced, stored in a retrieval system, or transmitted in any form or by any means, electronic, mechanical, photocopying, recording or otherwise, without the prior permission in writing of CITB-ConstructionSkills.

Produced by Thomson Prometric, printed in the UK

Contents

Contents

Foreword

The construction industry accounts for more than a quarter of all work-related deaths and over 4,000 major injuries each year. For this reason, health and safety continues to be at the forefront of the industry's agenda and is central to the drive to qualify the construction workforce by 2010. This commitment is also underpinned by legislation such as the Construction (Design and Management) regulations, which stipulate that firms must have skilled workers who understand the importance of health and safety.

The health and safety test plays an important role in meeting these objectives, and forms part of the requirement to obtain a CSCS or affiliated card. The test will continue to be a vital tool in helping the industry to improve health and safety standards.

To this end, we have been consulting with the industry and I am very pleased that over the coming years we will be introducing a significant set of improvements to the test. The test structure and question bank has already been revised and the test can be taken in the morning, afternoon and at the weekend – the times most convenient to people working in our industry.

CITB-ConstructionSkills is committed to continual improvement of delivery methods for specialist training and qualifications. We are currently working on behalf of industry, with the government and the HSE to drive forward improved standards.

If the whole industry works together on a number of small steps towards a fully competent and qualified workforce, we can have a huge impact and make it a safer place for all to work in.

Sir Michael Latham
Chairman

Preparing for the Health and Safety Test

Introduction

For many years the construction industry has had more than its fair share of serious accidents and deaths. Lack of training and little awareness of danger have been factors that have contributed to this unacceptable accident record. Too many accidents occur because the same mistakes are repeated over and over again.

The CITB-ConstructionSkills Health and Safety Test helps raise standards across the industry by ensuring that workers meet a minimum level of health and safety awareness before going on site. It now forms a key part of most major card schemes, including CSCS and its affiliates.

In April 2005, the test underwent significant improvements, with a revision of the question bank and an enhanced delivery infrastructure.

This guide has been written to help you prepare for the Health and Safety Test.

You will see that the guide is divided into three main areas:

- **Core test questions**
  Sections 1 – 15
- **Specialist test questions**
  Sections 16 – 21
- **HVACR test questions**
  Sections 22 – 26

Each of the 26 sections relates to a particular subject area and contains the actual test questions. All the questions that will make up your test paper are included in this guide, together with the answers.

Each CSCS (operatives) test will contain either 2 or 3 questions from each of the 15 Core sections with 40 questions in total.

For Supervisor tests, 20 questions from the Core sections will be asked, plus a further 20 questions from the Supervisory and Management section.

For Manager tests, 20 questions from the Core sections will be asked, plus a further 20 questions from the Supervisory and Management section. The Manager and Supervisor tests are mutually exclusive and results cannot be transferred between the two categories.

For the Specialist tests, 40 questions from the Core sections will be asked, plus a further 10 questions from the relevant Specialist section.

It is suggested that you read through this guide and attempt some of the questions from each section before attending the test. This will give you an idea of your strengths and weaknesses, and indicate the areas where you may need to improve your knowledge.

You may also want to prepare for your test by using a CITB-ConstructionSkills health and safety publication, which you can buy either when you book your test or online from www.citb-constructionskills.co.uk/publications

You can also attend one of the courses run by our National Construction College – call us on 08457 336666 to find out more. An official CD is also available which contains all the questions and can generate practice tests for each of the test categories.

It is NOT the purpose of the test to trip you up or to deliberately make things difficult for you. It is intended only to establish that you can demonstrate a satisfactory level of basic health and safety knowledge. There are no 'trick' questions.

The self-employed

Many of the questions will refer to the duties of employers. In law, the self-employed have the same legal responsibilities as employers. To keep the questions as brief as possible, the wording only refers to the duties of employers, and not to the duties of the self-employed, but the questions apply to both.

Your level of knowledge

You will NOT need a detailed knowledge of the exact content or wording of any regulations, but you will need to show that you know what is required of you, the things you must do (or not do) and what to do in certain circumstances, for example, if you discover an accident.

- **An example of what you must do** is to comply with the training you have been given on a piece of work equipment or machinery.
- **An example of what you must not do** is to damage or interfere with anything provided in the interests of health and safety, for example, damaging a fire extinguisher.
- **An example of what to do in certain circumstances** is to know what actions to take if you discover an accident or a fire.

All of the questions are set at this level, and if you are accustomed to taking care of your own health and safety, and that of others who may be affected by what you do, you should not have any difficulty in selecting the correct answers.

Northern Ireland legislation

All questions in this book are based on British legislation. However, Northern Ireland legislation differs from that in the rest of the UK.

Questions in Sections 1, 16 and 17 quoting legislation that is different in Northern Ireland are identified by an * after the question number (for example, 1.5 *). These questions are then adapted in the Appendix (at the back of the book) using Northern Ireland legislation.

For practical reasons, all candidates (including those in Northern Ireland) will be tested on questions using legislation relevant to the rest of the UK only.

The easiest ways to book your test are online or by telephone. You will be given the date and time of your test immediately. The test can also be booked by postal/fax application. You should expect that tests will be available 3 to 8 weeks from when you book.

The test is offered in over 150 Thomson Prometric test centres throughout the UK. The list of centres is available online at **www.citb-constructionskills.co.uk**

What you need to be able to make a booking

To make a test booking you will need

1. Your CITB-ConstructionSkills registration number – new applicants will be able to register by calling **0870 600 4020**.
   Existing cardholders can find this on their registration card or renewal letter.
2. To know which category of test you need to take.
3. A method of payment – credit or debit card if online, telephone, post or fax; cheque or postal order if sending application.

Online

You can make, check or change a test booking 24 hours a day, 7 days a week at **www.citb-constructionskills.co.uk**

You will receive an email confirming your test booking along with directions to the test centre. It is important that you check the details in this confirmation and follow any instructions it gives regarding the test.

Telephone

To book your test by telephone, call:

0870 600 4020

The registration centre is open between 8am and 6pm, Monday to Friday. All the details of your test will be confirmed with you during the call and again in a letter that you should receive within 10 days. It is important that you check the details in this confirmation and follow any instructions it gives regarding the test.

Booking your Health and Safety Test

By post/fax

If you prefer to book by post or fax you need to use the application form available at www.citbconstructionskills.co.uk or by calling 0870 600 4020. If you are making a booking for a group of candidates please make sure to ask for the **Group Booking** application form.

Please preferably fax the completed application form to:

0845 850 8012

or post to:

CITB-ConstructionSkills
PO Box 148
Salford
M5 3SY

Special needs

If you have any special requirements for taking the test, such as any reading difficulties or preferring to take the test in a language other than English, please advise us at the time of booking. Every test is available with English or Welsh voice over, i.e. candidates are given headphones so that all the questions and possible answers are 'read' to the candidate.

Corporate Mobile Testing and Internet Based Testing Centres

For organisations wishing to test a group of candidates at their place of work, please call 0161 868 3800 for details of the mobile testing service or becoming an internet based test centre.

What if I don't receive a confirmation letter?

If you do not receive a confirmation letter within the time specified, please telephone 0870 600 4020 to check that your test booking has been made.

We cannot take responsibility for postal delays. If you miss your test event, you will unfortunately forfeit your fee.

How do I cancel or postpone my test?

To cancel or postpone your test event you should go online or call the booking number at least three clear working days before your test date, otherwise you'll lose your fee.

Only in exceptional circumstances, such as documented ill health or family bereavement, can this rule be waived.

You must make sure that when you arrive at the test centre you have all the relevant documentation with you, or you will not be able to take your test and you'll lose your fee. The test centre will display both Thomson Prometric and CITB logos. Make sure you arrive in plenty of time so that you are not rushed. If you arrive after your session has started you may not be allowed to take the test.

You'll need:

- Your confirmation email or letter
- ID bearing your photo
- ID bearing your signature (this could be the same as your photo ID).

The test centre staff will check your documents and make sure that you take the category of test that you have booked.

The tests are all delivered on a computer screen. However you do not need to be familiar with computers to be able to pass the test, all you will need to do is click on the relevant answer boxes. The computer-based test does not involve any writing.

Before the test begins there will be an introduction and practice session to allow you to get used to the way the test will work.

The test will contain a number of multiple-choice questions. Most questions will give four answers, only one of which is correct. You must select the answer that you think is correct. Some questions will require you to select two correct answers from four or five, and others require three correct from a choice of five. Each question will be clearly marked with the number of correct answers that you must find.

For the CSCS operatives, managers and supervisors, the time allowed to answer the 40 questions is 45 minutes. For the specialist tests, the time allowed to answer the 50 questions is 60 minutes. A clock is always visible on screen to show the time remaining.

At the end of the test there is an optional survey which gives you the chance to provide feedback on the test process.

Remember

• no photo

• no signature

NO TEST

If you fail the test you will receive feedback on the subject areas in which you got questions wrong before leaving the test centre. If you do not reach the required standard you are strongly advised to read again all of the topic areas appropriate to your test.

Failure to pass the test shows that you're not fully prepared. You'll have to wait at least three clear working days before you take the test again.

Obtaining your CSCS/CPCS Card

It is becoming increasingly important to carry the right card to prove your skills on construction sites.

Once you have passed your Health and Safety Test, you should apply to join the appropriate card scheme for your trade. In most cases, this will be CSCS (the Construction Skills Certification Scheme), CPCS or an affiliated scheme.

Your pass letter should include contact details of all relevant schemes you can apply to and details on what to do next. You can also find out more about many of the schemes at:

www.citb-constructionskills.co.uk

Test Abbreviations

CSCS	Construction Skills Certification Scheme
DAP	Combined Demolition & Plant
DUCT	Ductwork (HVACR)
HSE	Health and Safety Executive
HAPS	Heating & Plumbing (HVACR)
HIW	Highways
LAEE	Lift and Escalator
LPG	Liquid Petroleum Gas
MGRS	Manager's test
MEWP	Mobile Elevating Work Platform
PPE	Personal Protective Equipment
PFW	Pipework (HVACR)
JIBP	Plumbing Health & Safety
RAAC	Refrigeration and Air Conditioning (HVACR)
RCD	Residual Current Device
RIDDOR	Reporting of Injuries, Diseases and Dangerous Occurrences Regulations
RPE	Respiratory Protective Equipment
SAF	Services and Facilities (HVACR)
SUP	Supervisory test
WAH	Working at Height

All candidates will be asked questions from the 'core' test sections, which include:

- Section 1 General Responsibilities
- Section 2 Accident Prevention and Reporting
- Section 3 Health and Welfare
- Section 4 Manual Handling
- Section 5 Working at Height
- Section 6 Personal Protective Equipment (PPE)
- Section 7 Emergency Procedures and First Aid
- Section 8 Safe Use of Hazardous Substances
- Section 9 Electrical Safety
- Section 10 Hand-held Equipment and Tools
- Section 11 Fire Prevention and Control
- Section 12 Safety Signs and Signals
- Section 13 Site Transport Safety
- Section 14 Noise and Vibration
- Section 15 Excavations and Confined Spaces

01 General Responsibilities

1.1

Who is responsible for reporting any unsafe conditions on site?

- [ ] A: The site manager only
- [ ] B: The client
- [ ] C: Health and Safety Executive inspectors
- [x] D: Everyone on site

1.2

A risk assessment tells you:

- [ ] A: how to report accidents
- [ ] B: the site working hours
- [x] C: how to do the job safely
- [ ] D: where the first-aid box is kept

1.3

During site induction, you do not understand something the presenter says. What should you do?

- [ ] A: Attend another site induction
- [x] B: Ask the presenter to explain the point again
- [ ] C: Guess what the presenter was trying to tell you
- [ ] D: Wait until the end then ask someone else to explain

1.4

Now that work on site is moving forward, the safety rules given in your site induction seem out of date. What should you do?

- [ ] A: Do nothing, you are not responsible for safety on site
- [x] B: Speak to your supervisor about your concerns
- [ ] C: Speak to your workmates to see if they have any new rules
- [ ] D: Make up your own safety rules to suit the changing conditions

1.5

The Health and Safety at Work Act places legal duties on:

- [ ] A: employers only
- [ ] B: operatives only
- [x] C: all people at work
- [ ] D: self-employed people only

1.6

As an employee, you do not have a legal duty to:

- [ ] A: use all equipment safely and as instructed
- [x] B: write your own risk assessments
- [ ] C: speak to your supervisor if you are worried about safety on site
- [ ] D: report any equipment that is damaged or defective

Answers: 1.1 = D, 1.2 = C, 1.3 = B, 1.4 = B, 1.5 = C, 1.6 = B

1.7

You will often hear the word hazard mentioned on site and during safety talks. What does it mean?

- [x] A: Anything at work that can harm you
- [ ] B: The site accident rate
- [ ] C: A type of barrier or machine guard
- [ ] D: All of the other answers

1.8

Why is the Health and Safety at Work Act important to you? Give two answers:

- [ ] A: it tells you which parts of the site are dangerous
- [ ] B: it must be learned before starting work
- [x] C: it requires your employer to provide a safe place to work
- [ ] D: it tells you how to do your job
- [x] E: it puts legal duties on you as an employee

1.9

It is important to see your employer's health and safety policy because it tells you:

- [ ] A: how to do your job safely
- [ ] B: how to write risk assessments
- [x] C: how health and safety is managed
- [ ] D: how to use tools and equipment safely

1.10

You are using some equipment. It has just been given a Prohibition Notice. What does this mean?

- [ ] A: You must not use it unless your supervisor is present
- [x] B: You must not use it until it is made safe
- [ ] C: You can use it as long as you take more care
- [ ] D: Only supervisors can use it

Answers: 1.7 = A, 1.8 = C,E, 1.9 = C, 1.10 = B

1.11

You are about to start a job. How will you know if it needs a Permit to Work?

- [ ] A: You will be given a Permit to Work at the site induction
- [ ] B: The Health and Safety Executive will tell you
- [ ] C: You will not be allowed to start work until the Permit to Work has been issued
- [x] D: You don't need to know, Permits to Work only affect managers

1.12

As an employee, which of these is not your duty under the Health and Safety at Work Act?

- [ ] A: To look after your own health and safety
- [ ] B: To look after the health and safety of anyone else who might be affected by your work
- [x] C: To write your own risk assessments
- [ ] D: Not to interfere with anything provided for health and safety

1.13

You find that you cannot do a job as the method statement says you should. What do you do?

- [ ] A: Make up your own way of doing the job
- [x] B: Do not start work until you have talked with your supervisor
- [ ] C: Ask your workmates how they think you should do the job
- [ ] D: Contact the Health and Safety Executive

1.14

Why do you need to know the health and safety regulations that deal with your type of work?

- [ ] A: They tell you how to write risk assessments
- [ ] B: They explain how health and safety is managed on your site
- [ ] C: They tell you when Health and Safety Executive inspections will take place
- [x] D: They place legal duties on you

Answers: 1.11 = C, 1.12 = C, 1.13 = B, 1.14 = D

1.15

The whole site has been issued with a Prohibition Notice. What does this mean to you?

- [ ] A: You must check with your supervisor before starting work
- [ ] B: You must check with the Health and Safety Executive before starting work
- [ ] C: You must not use any tools or machinery
- [x] D: You must stop work

1.16

After watching you work, a Health and Safety Executive inspector issues an Improvement Notice. What does this mean?

- [ ] A: You are not working fast enough
- [ ] B: You need to improve the standard of your work
- [x] C: You are not working in a safe way
- [ ] D: All of the other answers

8/9

1.17

You have been told that a Health and Safety Executive inspector is on site. What should you do?

- [ ] A: Stop work and report to your supervisor
- [ ] B: Stop work and go to the assembly point
- [x] C: Carry on working unless you are asked to stop
- [ ] D: Finish what you are doing and go home

1.18

Who is responsible for managing health and safety on construction sites?

- [ ] A: The police
- [ ] B: The Health and Safety Executive
- [ ] C: The client
- [x] D: The site manager

1.19

If you discover children playing on site, what is the first thing you should do?

- [ ] A: Tell your supervisor
- [ ] B: Phone the police
- [x] C: Make sure the children are taken to a place of safety
- [ ] D: Find out how they got into the site

Answers: 1.15 = D, 1.16 = C, 1.17 = C, 1.18 = D, 1.19 = C

1.20

What is a toolbox talk?

- [x] A: A short training session on a particular safety topic
- [ ] B: A talk that tells you where to buy tools
- [ ] C: Your first training session when you arrive on site
- [ ] D: A sales talk given by a tool supplier

1.21

When should you report environmental incidents and near misses?

- [ ] A: Never
- [ ] B: During your next break
- [x] C: As soon as practical
- [ ] D: At the end of the day

1.22

If you have any rubbish or leftover materials at the end of the job, you should:

- [ ] A: leave it where it is
- [ ] B: pick it up and throw it with rubbish left by other people
- [x] C: put it in the designated waste area
- [ ] D: pick it up and dump it outside the site

1.23

A Permit to Work allows:

- [ ] A: the emergency services to come on to the site after an accident
- [x] B: certain jobs to be carried out under controlled conditions
- [ ] C: Health and Safety Executive inspectors to visit the site
- [ ] D: untrained people to work without supervision

Answers: 1.20 = A, 1.21 = C, 1.22 = C, 1.23 = B

2.1

When must you record an accident in the accident book?

- [x] A: If you are injured in any way
- [ ] B: Only if you have to be off work
- [ ] C: Only if you have suffered a broken bone
- [ ] D: Only if you have to go to hospital

2.2

If someone is injured at work, who should record it in the accident book?

- [ ] A: The site manager, and no one else
- [x] B: The injured person, or someone acting for them
- [ ] C: The first-aider, and no one else
- [ ] D: Someone from the Health and Safety Executive

2.3

If you cut your finger and it won't stop bleeding, you should:

- [ ] A: Wrap something around it and carry on working
- [ ] B: Tell your supervisor
- [ ] C: Wash it clean then carry on working
- [x] D: Find a first-aider or get other medical help

10/10

2.4

Why is it important to report all accidents?

- [x] A: It might stop them happening again
- [x] B: Some types of accident have to be reported to the Health and Safety Executive
- [x] C: Details have to be entered in the accident book
- [x] D: All of the other answers

X

2.5

Which of these does not have to be recorded in the accident book?

- [x] A: Your national insurance number
- [ ] B: The date and time of your accident
- [ ] C: Details of your injury
- [ ] D: Your home address

2.6

A near miss accident is an accident where:

- [ ] A: you were just too late to see what happened
- [x] B: someone could have been injured
- [ ] C: someone was injured and nearly had to go to hospital
- [ ] D: someone was injured and nearly had to take time off work

Answers: 2.1 = A, 2.2 = B, 2.3 = D, 2.4 = D, 2.5 = A, 2.6 = B

2.7

Why is it important to report all near miss accidents?

- [ ] A: The Health and Safety Executive need to know about everything that happens on site
- [ ] B: To find someone to blame
- [ ] C: You might want to claim compensation
- [x] D: To learn from them and stop them happening again

2.8

A scaffold has collapsed and you saw it happen. When you are asked about the accident, you should say:

- [ ] A: nothing, you are not a scaffold expert
- [ ] B: as little as possible because you don't want to get people into trouble
- [x] C: exactly what you saw
- [ ] D: who you think is to blame and how they should be punished

2.9

You must remove nails from scrap wood because:

- [x] A: someone could stand on an upright nail and injure their foot
- [ ] B: scrap wood and scrap metal must be put in separate skips
- [ ] C: the site will run out of nails
- [ ] D: the wood will take up more room in the skip

2.10

You can help prevent accidents by:

- [x] A: reporting unsafe working conditions
- [ ] B: becoming a first-aider
- [ ] C: knowing where the first aid kit is kept
- [ ] D: knowing how to get help quickly

2.11

When are you most likely to have an accident?

- [ ] A: In the morning
- [ ] B: In the afternoon
- [ ] C: During the summer months
- [x] D: When you first start on site

2.12

What is the most important reason for keeping your work area clean and tidy?

- [x] A: To prevent slips, trips and falls
- [ ] B: So that you don't have a big clean-up at the end of the week
- [ ] C: So that waste skips can be emptied more often
- [ ] D: To recycle waste and help the environment

Answers: 2.7 = D, 2.8 = C, 2.9 = A, 2.10 = A, 2.11 = D, 2.12 = A

2.13

Which type of accident kills most construction workers?

- A: Falling from height
- B: Contact with electricity
- C: Being run over by site transport
- D: Being hit by a falling object

2.14

If you have a minor accident, who should report it?

- A: Anyone who saw the accident happen
- B: A subcontractor
- C: You
- D: The Health and Safety Executive

2.15

The work of another contractor is affecting your safety. You should stop work and:

- A: go home
- B: speak to your supervisor
- C: speak to the contractor's supervisor
- D: speak to the contractor who is doing the job

2.16

Which of these will help you to work safely on site?

- A: Site induction
- B: Toolbox talks
- C: Risk assessments and method statements
- D: All the other answers

2.17

How would you expect to find out about health and safety rules when you first arrive on site?

- A: During site induction
- B: In a letter sent to your home
- C: By reading your employer's health and safety policy
- D: By asking others on the site

2.18

If your doctor says that you have Weil's disease, you will need to tell your employer. Why?

- A: Your employer will not want to go anywhere near you
- B: Your employer will have to report it to the Health and Safety Executive
- C: Your workmates might catch it from you
- D: The whole site will have to be closed down

Answers: 2.13 = A, 2.14 = C, 2.15 = B, 2.16 = D, 2.17= A, 2.18 = B

2.19

Why is it important to attend site induction?

- [ ] A: You will get to know other new starters
- [ ] B: Risk assessments will be handed out
- [x] C: Site health and safety rules will be explained
- [ ] D: Permits to Work will be handed out

Answer: 2.19 = C

3.1

Two kinds of animal can carry a disease called Leptospirosis in their urine. Which two?

- A: Cat
- B: Sheep
- C: Rat
- D: Rabbit
- E: Cow

3.2

Look at these statements about illegal drugs in the workplace. Which one is true?

- A: Users of illegal drugs are a danger to everyone on site
- B: People who take illegal drugs work better and faster
- C: People who take illegal drugs take fewer days off work
- D: Taking illegal drugs is a personal choice so other people shouldn't worry about it

3.3

Direct sunlight on bare skin can cause:

- A: dermatitis
- B: rickets
- C: acne
- D: skin cancer

3.4

You should clean very dirty hands with:

- A: soap and water
- B: thinners
- C: white spirit
- D: paraffin

3.5

You need to handle a hazardous substance. You should wear the correct gloves to help stop:

- A: skin disease
- B: vibration white finger
- C: raynaud's syndrome
- D: arthritis

3.6

To help keep rats away, everyone on site should:

- A: buy rat traps and put them around the site
- B: ask the local authority to put down rat poison
- C: bring a large cat to site
- D: not leave scraps of food lying about

Answers: 3.1 = C,E, 3.2 = A, 3.3 = D, 3.4 = A, 3.5 = A, 3.6 = D

3.7

If you get a hazardous substance on your hands, it can pass from your hands to your mouth when you eat. Give two ways to stop this:

- [ ] A: wear protective gloves while you are working
- [ ] B: wash your hands before eating
- [ ] C: put barrier cream on your hands before eating
- [ ] D: wear protective gloves then turn them inside-out before eating
- [ ] E: wash your work gloves then put them on again before eating

3.8

It is your first day on site. You find that there is nowhere to wash your hands. What should you do?

- [ ] A: Wait until you get home then wash them
- [ ] B: Go to a local café or pub and use the washbasin in their toilet
- [ ] C: Speak to your supervisor about the problem
- [ ] D: Bring your own bottle of water the next day

3.9

What sort of rest area should your employer provide on site?

- [ ] A: A covered area
- [ ] B: A covered area and some chairs
- [ ] C: A covered area, tables and chairs, and something to heat water
- [ ] D: Nothing, employers don't have to provide rest areas

3.10

What is the minimum that should be provided on site for washing your hands?

- [ ] A: Nothing, there is no need to provide washing facilities
- [ ] B: Running hot water and electric hand-dryers
- [ ] C: A cold water standpipe and paper towels
- [ ] D: Hot and cold water (or warm water), soap and a way to dry your hands

Answers: 3.7 = A,B, 3.8 = C, 3.9 = C, 3.10 = D

3.11

Your doctor has given you some medication. Which of these questions is the most important?

- [x] A: Will it make me sleepy or unsafe to work?
- [ ] B: Will I work more slowly?
- [ ] C: Will my supervisor find out?
- [ ] D: Will I oversleep and be late for work?

3.12

You should not just rely on barrier cream to protect your skin from harmful substances because:

- [ ] A: it costs too much to use every day
- [x] B: many harmful substances go straight through it
- [ ] C: it is difficult to wash off
- [ ] D: it can irritate your skin

3.13

When should you use barrier cream?

- [x] A: Before you start work
- [ ] B: When you finish work
- [ ] C: As part of first aid treatment
- [ ] D: When you can't find your gloves

9/10

3.14

Someone goes to the pub at lunchtime and has a couple of pints of beer. What should they do next?

- [ ] A: Drink plenty of strong coffee then go back to work
- [x] B: Stay away from the site for the rest of the day
- [ ] C: Stay away for an hour and then go back to work
- [ ] D: Eat something, wait 30 minutes and then go back to work

3.15

Occupational asthma is a disease that can end your working life. It affects your:

- [ ] A: hearing
- [ ] B: joints
- [ ] C: skin
- [x] D: breathing

3.16

The site toilets do not flush. What should you do?

- [ ] A: Try not to use the toilets while you are at work
- [x] B: Tell your supervisor about the problem
- [ ] C: Try to fix the fault yourself
- [ ] D: Ask a plumber to fix the fault

Answers: 3.11 = A, 3.12 = B, 3.13 = A, 3.14= B, 3.15 = D, 3.16 = B

3.17

The toilets on your site are always dirty. What should you do?

- [ ] A: Ignore the problem, it is normal
- [x] B: Make sure that you tell someone who can sort it out
- [ ] C: Find some cleaning materials and do it yourself
- [ ] D: See if you can use the toilets in a nearby café or pub

3.18

You are more likely to catch Weil's disease (Leptospirosis) if you:

- [ ] A: work near wet ground, waterways or sewers
- [ ] B: work near air conditioning units
- [ ] C: fix showers or baths
- [x] D: drink water from a standpipe

3.19

Exposure to engine oil and other mineral oils can cause:

- [x] A: skin problems
- [ ] B: heart disease
- [ ] C: breathing problems
- [ ] D: vibration white finger

3.20

You can get occupational dermatitis from:

- [ ] A: hand-arm vibration
- [ ] B: another person with dermatitis
- [x] C: some types of strong chemical
- [ ] D: sunlight

3.21

There are many kinds of dust at work. Breathing them for a long time can cause:

- [x] A: occupational asthma
- [ ] B: occupational dermatitis
- [ ] C: skin cancer
- [ ] D: glue ear

3.22

You can catch an infection called tetanus from contaminated land or water. How does it get into your body?

- [ ] A: Through your nose when you breathe
- [x] B: Through an open cut in your skin
- [ ] C: Through your mouth when you eat or drink
- [ ] D: It doesn't, it only infects animals and not people

Answers: 3.17 = B, 3.18 = A, 3.19 = A, 3.20 = C, 3.21 = A, 3.22 = B

3.23

You should not use white spirit or other solvents to clean your hands because:

- [x] A: they strip the protective oils from the skin
- [ ] B: they remove the top layer of skin
- [ ] C: they block the pores of the skin
- [ ] D: they carry harmful bacteria that attack the skin

3.24

The early signs of Weil's disease (Leptospirosis) can be easily confused with:

- [ ] A: dermatitis
- [ ] B: diabetes
- [ ] C: hayfever
- [x] D: influenza ('flu')

Answers: 3.23 = A, 3.24 = D

04 Manual Handling

4.1

To lift a load, you should always try to:

- [ ] A: stand with your feet together when lifting
- [ ] B: bend your back when lifting
- [ ] C: carry the load away from your body, at arm's length
- [x] D: divide large loads into smaller loads

4.2

You need to lift a load from the floor. You should stand with your:

- [ ] A: feet together, legs straight
- [ ] B: feet together, knees bent
- [x] C: feet slightly apart, knees bent
- [ ] D: feet wide apart, legs straight

4.3

You have been told how to lift a heavy load, but you think there is a better way to do it. What should you do?

- [ ] A: Ignore what you have been told and do it your way
- [ ] B: Ask your workmates to decide which way you should do it
- [x] C: Discuss your idea with your supervisor
- [ ] D: Forget your idea and do it the way you have been told

4.4

To lift a load safely, you need to think about:

- [ ] A: its size and condition
- [ ] B: its weight
- [ ] C: whether it has handholds
- [x] D: all of the other answers

4.5

When you lift a load manually, you must:

- [x] A: keep your back straight and use the strength in your leg muscles to lift
- [ ] B: make sure there are always two people to lift the load
- [ ] C: use a crane or another lifting device to pick up the load
- [ ] D: move the load as quickly as possible

4.6

You are using a wheelbarrow to move a heavy load. Is this manual handling?

- [ ] A: No, because the wheelbarrow is carrying the load
- [ ] B: Only if the load slips off the wheelbarrow
- [x] C: Yes, you are still manually handling the load
- [ ] D: Only if the wheelbarrow has a flat tyre

Answers: 4.1 = D, 4.2 = C, 4.3 = C, 4.4 = D, 4.5 = A, 4.6 = C

4.7

You are using a trolley to move a heavy load. The trolley loses a wheel. You still have a long way to go. What should you do?

- [ ] A: Carry the load the rest of the way
- [ ] B: Ask someone to help you pull the trolley the rest of the way
- [ ] C: Drag the trolley on your own for the rest of the way
- [x] D: Find another way to move the load

4.8

You need to move a heavy load over a long distance. What is the safest way to do it?

- [ ] A: Pick up the load and run all the way
- [ ] B: Tie a rope around the load and pull it
- [ ] C: Roll it end-over-end all the way
- [x] D: Use a wheelbarrow or trolley

4.9

Your new job involves some manual handling. An old injury means that you have a weak back. What should you do?

- [ ] A: Tell your supervisor you can lift anything
- [x] B: Tell your supervisor that lifting might be a problem
- [ ] C: Try some lifting then tell your supervisor about your back
- [ ] D: Tell your supervisor about your back if it gets injured again

4.10

If you have to twist or turn your body when you lift and place a load, it means:

- [x] A: the weight you can lift safely is **less** than usual
- [ ] B: the weight you can lift safely is **more** than usual
- [ ] C: nothing, you can lift the **same** weight as usual
- [ ] D: you **must** wear a back brace

10/10

Answers: 4.7 = D, 4.8 = D, 4.9 = B, 4.10 = A

4.11

You have to move a load that might be too heavy for you. You cannot divide it into smaller parts and there is no one to help you. What should you do?

- [x] A: Do not move the load until you have found a safe method
- [ ] B: Get a forklift truck, even though you can't drive one
- [ ] C: Try to lift it using the correct lifting methods
- [ ] D: Lift and move the load quickly to avoid injury

4.12

You need to move a load that might be too heavy for you. What should you do?

- [ ] A: Divide the load into smaller loads if possible
- [ ] B: Get someone to help you
- [ ] C: Use an aid, such as a trolley or wheelbarrow
- [x] D: All of the other answers

4.13

You have to lift a heavy load. What must your employer do?

- [ ] A: Make sure your supervisor is there to advise while you lift
- [x] B: Do a risk assessment of the task
- [ ] C: Nothing, it is part of your job to lift loads
- [ ] D: Watch you while you lift the load

4.14

If you wear a back support belt:

- [ ] A: you can lift any load without being injured
- [ ] B: you can safely lift more than usual
- [x] C: you could face the same risk of injury as not wearing one
- [ ] D: it will crush your backbone and damage it

Answers: 4.11 = A, 4.12 = D, 4.13 = B, 4.14 = C

4.15

You need to lift a load that is not heavy, but it is so big that you cannot see in front of you. What should you do?

- [x] A: Ask someone to help carry the load so that you can both see ahead
- [ ] B: Get someone to walk next to you and give directions
- [ ] C: Get someone to walk in front of you and tell others to get out of the way
- [ ] D: Move the load on your own because it is so large that anyone in your way is sure to see it

4.16

You need to move a load that is heavier on one side than the other. How should you pick it up?

- [x] A: With the heavy side towards you
- [ ] B: With the heavy side away from you
- [ ] C: With the heavy side on your strong arm
- [ ] D: With the heavy side on your weak arm

5/5

4.17

Someone is going to help you to lift a load. It is important that both of you:

- [ ] A: work for the same employer
- [x] B: are about the same size and can lift the same weight
- [ ] C: are about the same age
- [ ] D: are right-handed (or both left-handed)

4.18

You have to move a load while you are sitting, not standing. How much can you lift safely?

- [x] A: Less than usual
- [ ] B: The usual amount
- [ ] C: Twice the usual amount
- [ ] D: Three times the usual amount

4.19

You need to reach above your head and lower a load to the floor. Which of these is not true?

- [ ] A: It will be more difficult to keep your back straight and chin tucked in
- [ ] B: You will put extra stress on your arms and your back
- [x] C: You can safely handle more weight than usual
- [ ] D: The load will be more difficult to control

Answers: 4.15 = A, 4.16 = A, 4.17 = B, 4.18 = A, 4.19 = C

4.20

Which part of your body is most likely to be injured if you lift heavy loads?

- [ ] A: Your knees
- [x] B: Your back
- [ ] C: Your shoulders
- [ ] D: Your elbows

4.21

Who should decide what weight you can lift safely?

- [x] A: You
- [ ] B: Your supervisor
- [ ] C: Your employer
- [ ] D: The Health and Safety Executive

4.22

What is the safest way to find out if a load is too heavy to lift?

- [ ] A: Pick it up quickly then put it down
- [ ] B: Walk round the load, and look at it from all sides
- [x] C: Find out the weight of the load
- [ ] D: See if you can hold the load at arm's length

4.23

You have to carry a load down a steep slope. What should you do?

- [ ] A: Walk backwards down the slope to improve your balance
- [ ] B: Carry the load on your shoulder
- [x] C: Assess whether you can still carry the load safely
- [ ] D: Run down the slope to finish quickly

4.24

Under the regulations for manual handling, all employees must:

- [ ] A: wear back-support belts when lifting anything
- [ ] B: make a list of all the heavy things they have to carry
- [ ] C: lift any size of load once the risk assessment has been done
- [x] D: make full use of their employer's safe systems of work

5/5

Answers: 4.20 = B, 4.21 = A, 4.22 = C, 4.23 = C, 4.24 = D

5.1

Who should erect, dismantle or alter a tube and fitting scaffold?

- A: Anyone who thinks they can do it
- B: Anyone who has the right tools
- C: Anyone who is competent and authorised
- D: Anyone who is a project manager

5.2

It is safe to cross a fragile roof if you:

- A: walk along the line of bolts
- B: can see fragile roof signs
- C: don't walk on any plastic panels
- D: use crawling boards

5.3

You are working on a flat roof. What is the best way to stop yourself falling over the edge?

- A: Put a large warning sign at the edge of the roof
- B: Ask someone to watch you and shout when you get too close to the edge
- C: Protect the edge with a guard-rail and toe-board
- D: Use red and white tape to mark the edge

5.4

A ladder should not be painted because:

- A: the paint will make it slippery to use
- B: the paint may hide any damaged parts
- C: the paint could damage the metal parts of the ladder
- D: it will need regular re-painting

5.5

How many people should be on a ladder at the same time?

- A: 2
- B: 1
- C: 1 on each section of an extension ladder
- D: 3 if it is long enough

5.6

You find a ladder that is damaged. What should you do?

- A: Don't use it and make sure that others know about the damage
- B: Don't use it and report the damage at the end of your shift
- C: Try and mend the damage
- D: Use the ladder if you can avoid the damaged part

6/6

Answers: 5.1 = C, 5.2 = D, 5.3 = C, 5.4 = B, 5.5 = B, 5.6 = A

5.7

You need to use a ladder to reach a work platform. What should be the slope or angle of the ladder?

- [ ] A: 45°
- [x] B: 60°
- [ ] C: 75°
- [ ] D: 85°

5.8

You need to stack materials on a scaffold platform. What is the best way to stop them falling over the toe-board?

- [ ] A: Fit brick guards
- [ ] B: Put a warning sign on the stack
- [x] C: Build the stack so that it leans away from the edge
- [ ] D: Cover the stack with netting

5.9

A scaffold guard-rail must be removed to allow materials to be lifted onto the platform. You are not a scaffolder. Can you remove the guard-rail?

- [x] A: Yes, if you put it back as soon as the load has been landed
- [ ] B: Yes, if you put it back before the end of your shift
- [ ] C: No, only a scaffolder can remove the guard-rail but you can put it back
- [ ] D: No, only a scaffolder can remove the guard-rail and put it back

5.10

Who should check a ladder before it is used?

- [x] A: The person who is going to use it
- [ ] B: A supervisor
- [ ] C: The site safety officer
- [ ] D: The manufacturer

1/4

Answers: 5.7 = C, 5.8 = A, 5.9 = D, 5.10 = A

5.11

What is the best way to make sure that a ladder is secure and won't slip?

- [ ] A: Tie it at the top
- [ ] B: Ask someone to stand with their foot on the bottom rung
- [x] C: Tie it at the bottom
- [ ] D: Wedge the bottom of the ladder with blocks of wood

5.12

You are working above water and there is a risk of falling. Which two items of Personal Protective Equipment do you need?

- [ ] A: Wellington boots
- [x] B: Harness and lanyard
- [x] C: Life jacket
- [ ] D: Waterproof jacket
- [ ] E: Waterproof trousers

3/4

5.13

You need to use a ladder to get to a scaffold platform. Which of these statements is true?

- [x] A: It must be tied and extend about five rungs above the platform
- [ ] B: All broken rungs must be clearly marked
- [ ] C: It must be wedged at the bottom to stop it slipping
- [ ] D: Two people must be on the ladder at all times

5.14

Tools and materials can easily fall from a scaffold platform. What is the best way to protect the people below?

- [ ] A: Make sure they are wearing safety helmets
- [ ] B: Tell them you will be working above them
- [x] C: Use brickguards to stop any items falling below
- [ ] D: Tell the people below to stop work and clear the area

Answers: 5.11 = A, 5.12 = B, C, 5.13 = A, 5.14 = C

5.15

When can you use a ladder as a place of work?

- [ ] A: If it is long enough
- [ ] B: If you can find a ladder to use
- [ ] C: If other people do not need to use it for access
- [x] D: If you are doing light work for a short time

5.16

You need to work at height. It is not possible to install edge protection or a soft landing system. What should you do?

- [ ] A: Hold onto something while you use your other hand to do the work
- [ ] B: Ask someone to hold you while you work
- [x] C: Wear a harness and lanyard and fix it to an anchor point
- [ ] D: Tie a rope round your waist and tie the other end to an anchor point

5.17

What is the best way to stop people falling through fragile roof panels?

- [ ] A: Tell everyone where the panels are
- [x] B: Cover the panels with something that can take the weight of a person
- [ ] C: Cover the panels with netting
- [ ] D: Mark the panels with red and white tape

5.18

You need to use a mobile tower scaffold. The wheel brakes do not work. What should you do?

- [ ] A: Use some wood to wedge the wheels and stop them moving
- [x] B: Do not use the tower
- [ ] C: Only use the tower if the floor is level
- [ ] D: Get someone to hold the tower while you use it

5.19

When you climb a ladder, you must:

- [x] A: have three points of contact with the ladder at all times
- [ ] B: have two points of contact with the ladder at all times
- [ ] C: use a safety harness
- [ ] D: have two people on the ladder at all times

Answers: 5.15 = D, 5.16 = C, 5.17 = B, 5.18 = B, 5.19 = A

5.20

What does this sign mean?

- [ ] A: Do not run on the roof
- [ ] B: Slippery when wet
- [x] C: Fragile roof
- [ ] D: Load-bearing roof

5.21

If you store materials on a working platform, you must make sure:

- [ ] A: the materials are secure, even in windy weather
- [ ] B: the platform can take the weight of the materials
- [ ] C: the materials do not make the platform unsafe for others
- [x] D: all of the other answers

5.22

You need to reach the working platform of a mobile tower scaffold. What is the right way to do this?

- [ ] A: Climb up the tower frame on the outside of the tower
- [ ] B: Lean a ladder against the tower and climb up that
- [x] C: Climb up the ladder built into the tower
- [ ] D: Jump from the rigid structure on which you are working

5.23

A mobile tower scaffold must not be used on:

- [x] A: soft or uneven ground
- [ ] B: a paved patio
- [ ] C: an asphalt road
- [ ] D: a smooth concrete path

5.24

You are working at height when you could fall from:

- [ ] A: the first lift of a scaffold or higher
- [ ] B: 2 metres above the ground or higher
- [x] C: any height that would cause an injury if you fell
- [ ] D: 3 metres above the ground or higher

10/10

Answers: 5.20 = C, 5.21 = D, 5.22 = C, 5.23 = A, 5.24 = C

Personal Protective Equipment (PPE)

6.1

You must wear head protection on site at all times unless you are:

- A: self-employed
- B: working alone
- C: in a safe area, like the site office
- D: working in very hot weather

6.2

Your employer must supply you with Personal Protective Equipment:

- A: twice a year
- B: if you pay for it
- C: if it is in the contract
- D: if you need to be protected

6.3

Do you have to pay for any Personal Protective Equipment you need?

- A: Yes, you must pay for all of it
- B: Only if you need to replace lost or damaged Personal Protective Equipment
- C: Yes, but you only have to pay half the cost
- D: No, your employer must pay for it

6.4

When should you wear safety boots or safety shoes on site?

- A: Only when you work at ground level
- B: In the winter
- C: Only when it is cold and wet
- D: All the time

6.5

To get the maximum protection from your safety helmet, you should wear it:

- A: back to front, to stop the peak banging into things
- B: pushed back on your head, to see better
- C: square on your head, to stop it falling off
- D: pulled forward, to protect your eyes

6.6

Who should provide you with any Personal Protective Equipment you need?

- A: Your employer
- B: You must buy your own
- C: Anyone on site with a budget
- D: No one has a duty to provide it

6/6

Answers: 6.1 = C, 6.2 = D, 6.3 = D, 6.4 = D, 6.5 = C, 6.6 = A

6.7

If your Personal Protective Equipment gets damaged, you should:

- [ ] A: throw it away and work without it
- [x] B: stop what you are doing until it is replaced
- [ ] C: carry on wearing it but work more quickly
- [ ] D: try to repair it

6.8

Look at these statements about wearing a safety helmet in hot weather. Which one is true?

- [ ] A: You can drill holes in it to keep your head cool
- [ ] B: You can wear it back-to-front if it is more comfortable that way
- [ ] C: You must take it off during the hottest part of the day
- [x] D: You must wear it at all times and in the right way

5/5

6.9

You have to work outdoors in bad weather. Your employer should supply you with waterproof clothing because:

- [ ] A: it will have the company name and logo on it
- [x] B: you are less likely to get muscle strains if you are warm and dry
- [ ] C: you are less likely to catch Weil's disease if you are warm and dry
- [ ] D: your supervisor will be able to see you more clearly in the rain

6.10

When do you need to wear eye protection?

- [ ] A: On very bright, sunny days
- [x] B: If there is a risk of eye injury
- [ ] C: When your employer can afford it
- [ ] D: Only if you work with chemicals

6.11

If there is a risk of materials flying into your eyes, you should wear:

- [ ] A: tinted welding goggles
- [ ] B: laser safety glasses
- [ ] C: chemical-resistant goggles
- [x] D: impact-resistant goggles

Answers: 6.7 = B, 6.8 = D, 6.9 = B, 6.10= B, 6.11 = D

6.12

Look at these statements about Personal Protective Equipment. Which one is not true?

- [x] A: You must pay for any damage or loss
- [ ] B: You must store it correctly when you are not using it
- [ ] C: You must report any damage or loss to your supervisor
- [ ] D: You must use it as instructed

6.13

Do all types of glove protect your hands against chemicals?

- [ ] A: Yes, all gloves are made to the same standard
- [ ] B: Only if you put barrier cream on your hands as well
- [x] C: No, different types of glove protect against different types of hazard
- [ ] D: Only if you cover the gloves with barrier cream

6.14

If you drop your safety helmet from height on to a hard surface, you should:

- [ ] A: repair any cracks then carry on wearing it
- [ ] B: make sure there are no cracks then carry on wearing it
- [ ] C: work without a safety helmet until you can get a new one
- [x] D: stop work and get a new safety helmet

6.15

You need to wear a full body harness. You have never used one before. What should you do?

- [x] A: Ask for expert advice and training
- [ ] B: Ask someone already wearing a harness to show you what to do
- [ ] C: Try to work it out for yourself
- [ ] D: Read the instruction book

Answers: 6.12 = A, 6.13 = C, 6.14 = D, 6.15 = A

6.16

You need special Respiratory Protective Equipment to handle a chemical. None has been provided. What should you do?

- [ ] A: Get on with the job but try to work quickly
- [x] B: Do not start work until you have been given the correct Respiratory Protective Equipment and training
- [ ] C: Start the work but take a break now and again
- [ ] D: Sniff the substance to see if it makes you feel ill

6.17

You have been given a dust mask to protect you against hazardous fumes. What should you do?

- [x] A: Do not start work until you have the correct Respiratory Protective Equipment
- [ ] B: Do the job but work quickly
- [ ] C: Start work but take a break now and again
- [ ] D: Wear a second dust mask on top of the first one

3/4

6.18

Look at these statements about anti-vibration gloves. Which one is true?

- [x] A: They might not protect you against vibration
- [ ] B: They cut out all hand-arm vibration
- [x] C: They only work against low frequency vibration
- [ ] D: They give the most protection if they are worn over other gloves

6.19

You are about to start a job. How will you know if you need any extra Personal Protective Equipment?

- [ ] A: By looking at your employer's health and safety policy
- [ ] B: You will just be expected to know
- [x] C: From the risk assessment or method statement
- [ ] D: A letter will be sent to your home

Answers: 6.16 = B, 6.17 = A, 6.18 = A, 6.19 = C

6.20

You have been given disposable ear-plugs to use, but they keep falling out. What should you do?

- [ ] A: Throw them away and work without them
- [x] B: Stop work until you get more suitable ones and are shown how to fit them
- [ ] C: Put two ear plugs in each ear so they stay in place
- [ ] D: Put rolled-up tissue paper in each ear

Answer: 6.20 = B

7.1

What is the one thing a first aider cannot do for you?

- [ ] A: Give mouth-to-mouth resuscitation
- [ ] B: Stop any bleeding
- [x] C: Give you medicines without authorisation
- [ ] D: Treat you if you are unconscious

7.2

A first aid box should not contain:

- [ ] A: bandages
- [ ] B: plasters
- [ ] C: safety pins
- [x] D: pain killers

7.3

If you want to be a first aider, you should:

- [ ] A: watch a first aider treating people then try it yourself
- [x] B: ask if you can do a first aider's course
- [ ] C: buy a book on first aid and start treating people
- [ ] D: speak to your doctor about it

7.4

What is the first thing you should do if you find an injured person?

- [ ] A: Tell your supervisor
- [x] B: Check that you are not in any danger
- [ ] C: Move the injured person to a safe place
- [ ] D: Ask the injured person what happened

7.5

If you think someone has a broken leg, you should:

- [ ] A: lie them on their side in the recovery position
- [ ] B: use your belt to strap their legs together
- [x] C: send for the first aider or get other help
- [ ] D: lie them on their back

7.6

If someone gets some grit in their eye, the best thing you can do is:

- [ ] A: hold the eye open and wipe it with clean tissue paper
- [ ] B: ask them to rub the eye until it starts to water
- [ ] C: tell them to blink a couple of times
- [x] D: hold the eye open and flush it with clean water

6/6

Answers: 7.1 = C, 7.2 = D, 7.3 = B, 7.4 = B, 7.5 = C, 7.6 = D

Emergency Procedures and First Aid

7.7

Someone gets a large splinter in their hand. It is deep under the skin and it hurts. What should you do?

- [ ] A: Use something sharp to dig it out
- [x] B: Make sure they get first aid
- [ ] C: Tell them to ignore it and let the splinter come out on its own
- [ ] D: Try to squeeze out the splinter with your thumbs

7.8

Someone collapses with stomach pain. There is no first aider on site. What should you do?

- [ ] A: Get them to sit down
- [x] B: Get someone to call the emergency services
- [ ] C: Get them to lie down in the recovery position
- [ ] D: Give them some pain killers

7.9

If someone falls and is knocked unconscious, you should:

- [ ] A: turn them over so they are lying on their back
- [x] B: send for medical help
- [ ] C: slap their face to wake them up
- [ ] D: give mouth-to-mouth resuscitation

7.10

Someone has fallen from height and has no feeling in their legs. You should tell them to:

- [ ] A: roll onto their back and keep their legs straight
- [ ] B: roll on to their side and bend their legs
- [x] C: stay where they are until medical help arrives
- [ ] D: raise their legs to see if any feeling comes back

7.11

Someone has got a nail in their foot. You are not a first aider. You must not pull out the nail because:

- [ ] A: you will let air and bacteria get into the wound
- [x] B: the nail is helping to reduce the bleeding
- [ ] C: it will prove that the casualty was not wearing safety boots
- [ ] D: the nail is helping to keep their boot on

7.12

This sign means:

- [ ] A: wear eye protection
- [x] B: eye-wash station
- [ ] C: risk of splashing
- [ ] D: shower block

Answers: 7.7 = B, 7.8 = B, 7.9 = B, 7.10 = C, 7.11 = B, 7.12 = B

7.13

This sign means:

- [x] A: first aid
- [ ] B: safe to cross
- [ ] C: no waiting
- [ ] D: wait here for help

7.14

You will find out about emergency assembly points from:

- [ ] A: a risk assessment
- [ ] B a method statement
- [x] C: the site induction
- [ ] D: the Permit to Work

7.15

If someone burns their hand, the best thing you can do is:

- [x] A: put the hand into cold water
- [ ] B tell them to carry on working to exercise the hand
- [ ] C: rub barrier cream or Vaseline into the burn
- [ ] D: wrap your handkerchief around the burn

12/12

7.16

You have to work alone on a remote part of the site. What would you expect to be given?

- [x] A: A small first aid kit
- [ ] B: The first aid box out of the office
- [ ] C: Nothing
- [ ] D: A book on first aid

7.17

Someone working in a deep manhole has collapsed. What is the first thing you should do?

- [ ] A: Get someone to lower you into the manhole on a rope
- [ ] B Climb into the manhole and give mouth-to-mouth resuscitation
- [ ] C: Go and tell your supervisor
- [x] D: Shout to let others know what has happened

7.18

The first aid box on site is always empty. What should you do?

- [ ] A: Bring your own first aid supplies into work
- [ ] B Find out who is taking all the first aid supplies
- [x] C: Find out who looks after the first aid box and let them know
- [ ] D: Ignore the problem, it is always the same

Answers: 7.13 = A, 7.14 = C, 7.15 = A, 7.16 = A, 7.17 = D, 7.18 = C

7.19

Does your employer have to provide a first aid box?

- [x] A: Yes, every site must have one
- [ ] B: **Only** if more than 50 people work on site
- [ ] C: **Only** if more than 25 people work on site
- [ ] D: No, there is no legal duty to provide one

7.20

When would you expect eye-wash bottles to be provided?

- [ ] A: **Only** on demolition sites where asbestos has to be removed
- [ ] B: **Only** on sites where refurbishment is being carried out
- [x] C: On all sites where people could get something in their eyes
- [ ] D: On all sites where showers are needed

5/5

7.21

How can you find out the emergency telephone number for your site? Give two answers.

- [x] A: Attend the site induction
- [x] B: Read the site notice boards
- [ ] C: Ask the Health and Safety Executive
- [ ] D: Ask the local hospital
- [ ] E: Look in the BT telephone directory

7.22

If there is an emergency on site, you should:

- [ ] A: leave the site and go home
- [ ] B: phone home
- [x] C: follow the site emergency procedure
- [ ] D: phone the Health and Safety Executive

7.23

If someone is in contact with a live cable, the best thing you can do is:

- [ ] A: phone the electricity company
- [ ] B: dial 999 and ask for an ambulance
- [x] C: switch off the power and call for help
- [ ] D: pull them away from the cable

Answers: 7.19 = A, 7.20 = C, 7.21 = A, B, 7.22 = C, 7.23 = C

7.24

This sign means:

- [ ] A: one-way system
- [ ] B: public right of way
- [x] C: assembly point
- [ ] D: site transport route

Answer: 7.24 = C

Safe Use of Hazardous Substances

8.1

The COSHH regulations deal with:

- [ ] A: the safe use of tools and equipment
- [ ] B: the safe use of lifting equipment
- [x] C: the safe use of hazardous substances
- [ ] D: safe working at height

8.2

Which of these statements about asbestos is true?

- [ ] A: Brown asbestos is safe but blue asbestos is a hazard to health
- [ ] B: White asbestos is safe to use
- [ ] C: All types of asbestos are safe to handle
- [x] D: All types of asbestos are a hazard to health

8.3

If you think you have found some asbestos, the first thing you should do is:

- [x] A: stop work and warn others
- [ ] B: take a sample to your supervisor
- [ ] C: put the bits in a bin and carry on with your work
- [ ] D: find the first aider

8.4

This symbol tells you a substance is:

- [ ] A: harmful
- [x] B: toxic
- [ ] C: corrosive
- [ ] D: an irritant

8.5

This symbol tells you a substance is:

- [ ] A: harmful
- [ ] B: toxic
- [x] C: corrosive
- [ ] D: an irritant

8.6

Which symbol means corrosive substance?

- [ ] A:
- [x] B:
- [ ] C:
- [ ] D:

6/6

Answers: 8.1 = C, 8.2 = D, 8.3 = A, 8.4 = B, 8.5 = C, D, 8.6 = B

8.7

If a substance has this symbol, why must you take care? Give two answers.

A: It can catch fire easily

B: It can irritate your skin

C: It can harm your health

D: It can kill you

E: It can burn your skin

8.8

You have to use a harmful substance. What must your supervisor do?

A: Let you get on without giving any instructions

B: Make sure that someone is working close to you

C: Watch while you use the substance

D: Tell you what is in the COSHH assessment

8.9

Which symbol means toxic substance?

A:

B:

C:

D:

8.10

Which symbol means harmful substance?

A:

B:

C:

D:

8.11

You find a bottle of chemicals. The bottle does not have a label. What is the first thing you should do?

A: Smell the chemical to see what it is

B: Put it in a bin to get rid of it

C: Put it somewhere safe then report it

D: Taste the chemical to see what it is

8.12

How can you tell if a product is hazardous?

A: By a symbol on the container label

B: By the shape of the container

C: It will always be in a black container

D: It will always be in a cardboard box

Answers: 8.7 = B,C , 8.8 = D, 8.9 = A, 8.10 = C, 8.11 = C , 8.12 = A

8.13

Which of these does not cause skin problems?

- [ ] A: Asbestos
- [ ] B: Bitumens
- [ ] C: Epoxy resins
- [ ] D: Solvents

8.14

Which of these will give you health and safety information about a hazardous substance?

- [ ] A: The site diary
- [ ] B: The delivery note
- [ ] C: The COSHH assessment
- [ ] D: The accident book

8.15

You need to use a hazardous substance. Who should explain the COSHH assessment before you start?

- [ ] A: An Health and Safety Executive inspector
- [ ] B: The site first aider
- [ ] C: Your supervisor
- [ ] D: The site security people

8.16

A COSHH assessment tells you how:

- [ ] A: to lift heavy loads and how to protect yourself
- [ ] B to work safely in confined spaces
- [ ] C: a substance might harm you and how to protect yourself
- [ ] D: noise levels are assessed and how to protect your hearing

8.17

If a substance has this symbol, you must take care because it can:

- [ ] A: burn your skin
- [ ] B: kill you
- [ ] C: catch fire easily
- [ ] D: irritate your skin

8.18

If a substance has this symbol, you must take care because it can:

- [ ] A: kill you
- [ ] B: cause a mild skin rash
- [ ] C: burn your skin
- [ ] D: catch fire easily

Answers: 8.13 = A, 8.14 = C, 8.15= C, 8.16= C, 8.17= B, 8.18 = C

8.19

Where should this label be used?

- A: Anywhere that asbestos has been used
- B: Fixed to pipes lagged with asbestos
- C: Fixed to bags of asbestos waste
- D: On new materials that contain asbestos

8.20

If you breathe in asbestos dust, it can cause:

- A: aching muscles
- B: influenza ('flu)
- C: lung disease
- D: painful joints

8.21

The *safest* way to use a hazardous substance is to:

- A: get on with the job as quickly as possible
- B: read your employer's health and safety policy
- C: read the COSHH assessment and follow the instructions
- D: ask someone who has already used it

8.22

You are on site. You need to throw away some waste liquid that has oil in it. What should you do?

- A: Pour it down a drain outside the site
- B: Pour it onto the ground and let it soak away
- C: Use it to start a fire
- D: Find out how you should get rid of it

8.23

How should you get rid of hazardous waste?

- A: Put it in any skip on site
- B: In accordance with the site rules
- C: Bury it on site
- D: Take it to the nearest local authority waste tip

Answers: 8.19 = C, 8.20= C, 8.21 = C, 8.22 = D, 8.23 = B

Electrical Safety

9.1

On building sites, the recommended safe voltage for electrical equipment is:

- [ ] A: 12 volts
- [ ] B: 24 volts
- [ ] C: 110 volts
- [ ] D: 230 volts

9.2

It is safe to work close to an overhead power line if:

- [ ] A: you do not touch the line for more than 30 seconds
- [ ] B: you use a wooden ladder
- [ ] C: the power is switched off
- [ ] D: it is not raining

9.3

This warning sign means:

- [ ] A: risk of electrocution
- [ ] B: risk of thunder
- [ ] C: electrical appliance
- [ ] D: risk of lightning

9.4

You are using a 230 volt item of equipment when the fuse blows. What is the first thing you should do?

- [ ] A: Fit another fuse of the same size
- [ ] B: Fit a bigger fuse
- [ ] C: Use a paperclip to link the contacts
- [ ] D: Check for obvious damage

9.5

When do you need to check electrical hand tools for damage?

- [ ] A: Before you use it
- [ ] B: Every day
- [ ] C: Once a week
- [ ] D: At least once a year

9.6

You need to use an extension cable. What two things must you do?

- [ ] A: Only uncoil the length of cable you need
- [ ] B: Uncoil the whole cable
- [ ] C: Clean the whole cable with a damp cloth
- [ ] D: Check the whole cable for damage
- [ ] E: Only check the cable you need for damage

Answers: 9.1 = C, 9.2 = C, 9.3 = A, 9.4 = D, 9.5 = A, 9.6 = B,D,

9.7

You should use an Residual Current Device with 230 volt tools because:

- [ ] A: it lowers the voltage
- [x] B: it quickly cuts off the power if there is a fault
- [ ] C: it makes the tool run at a safe speed
- [ ] D: it saves energy and lowers costs

9.8

The colour of a 110 volt power cable and connector should be:

- [ ] A: black
- [x] B: red
- [ ] C: blue
- [ ] D: yellow

9.9

You are using an electric drill when it cuts out. You should:

- [ ] A: shake it to see if it will start again
- [ ] B: pull the electric cable to see if it is loose
- [ ] C: switch the power off and on a few times
- [x] D: switch off the power and look for signs of damage

9.10

There is smoke coming from the motor of your electric drill. You should:

- [ ] A: pour water over it
- [ ] B: use a carbon dioxide extinguisher
- [x] C: unplug the drill and see that no one else uses it
- [ ] D: allow the drill to cool for 30 minutes then try again

9.11

You need to run an electrical cable across an area used by vehicles. What two things should you do?

- [ ] A: Wrap the cable in yellow tape so that drivers can see it
- [x] B: Cover the cable with a protection ramp
- [ ] C: Cover the cable with scaffold boards
- [x] D: Put up a sign that says 'Ramp Ahead'
- [ ] E: Run the cable at head height

Answers: 9.7 = B, 9.8 = D, 9.9 = D, 9.10 = C, 9.11 = B,D,

9.12

You need to work near an electrical cable. The cable has bare wires. What should you do?

- [ ] A: Quickly touch the cable to see if it is live
- [ ] B: Check there are no sparks coming from the cable and then start work
- [x] C: Tell your supervisor and keep well away
- [ ] D: Push the cable out of the way so that you can start work

9.13

If an extension cable has a cut in its outer cover, you should:

- [ ] A: check the copper wires don't show through the cut then use the cable
- [ ] B: put electrical tape around the damaged part
- [x] C: report the fault and make sure no one else uses the cable
- [ ] D: put a bigger fuse in the cable plug

4/5

9.14

What is the best way to protect an extension cable while you work?

- [x] A: Run the cable above head height
- [x] B: Run the cable by the shortest route
- [ ] C: Cover the cable with yellow tape
- [ ] D: Cover the cable with pieces of wood

9.15

How do you check if the Residual Current Device connected to a power tool is working?

- [ ] A: Switch the tool on and off
- [x] B: Press the test button on the Residual Current Device unit
- [ ] C: Switch the power on and off
- [ ] D: Run the tool at top speed to see if it cuts out

9.16

If you see burn marks on the casing of an electric drill, it means the drill has:

- [ ] A: been held too tightly
- [x] B: had an electrical fault
- [ ] C: been left in the sun
- [ ] D: been left in the rain

Answers: 9.12 = C, 9.13 = C, 9.14 = A, 9.15 = B, 9.16 = B

9.17

The PAT test label on a power tool tells you:

- A: when the next safety check is due
- B: when the tool was made
- C: who tested the tool before it left the factory
- D: its earth-loop impedance

9.18

The temporary 110 volt electrical distribution box you want to use is too far away. What should you do?

- A: Unplug the other extension leads and move the distribution box yourself
- B: Tell an electrician who is working nearby to move it for you
- C: Ask the supervisor to arrange for it to be moved
- D: Use several extension leads plugged into each other

9.19

Why should you try to use battery-powered tools rather than electrical ones?

- A: They are cheaper to run
- B: They will not give you an electric shock
- C: They will not give you hand-arm vibration
- D: They do not need to be tested or serviced

9.20

Why do building sites use a 110 volt electricity supply instead of the usual 230 volt domestic supply?

- A: It is cheaper
- B: It is less likely to kill you
- C: It moves faster along the cables
- D: It is safer for the environment

9.21

If you need to use a power tool in a waterlogged part of the site, it is safest to:

- A: wear Wellington boots
- B: use an air-powered tool if possible
- C: only use 230 volt equipment
- D: wrap a plastic bag around the tool

9.22

You need to use a 230 volt item of equipment. How should you protect yourself from an electric shock?

- A: Use a generator
- B: Put up safety screens around you
- C: Use a portable Residual Current Device
- D: Wear rubber boots and gloves

Answers: 9.17 = A, 9.18 = C, 9.19 = B, 9.20 = B, 9.21 = B, 9.22 = C

Hand-held Equipment and Tools

10.1

To operate a powered hand tool, you must be:

- [ ] A: over 16 years old
- [ ] B: over 18 years old
- [x] C: trained and competent
- [ ] D 21 years old or over

10.2

If you need to use a hand tool or power tool on site, it must be:

- [ ] A: made in the UK
- [ ] B: the right tool for the job and inspected at the start of each week
- [ ] C: bought from a builders' merchant
- [x] D: the right tool for the job and inspected before you use it

10.3

Before you adjust an electric hand tool, you should:

- [ ] A: switch it off but leave the plug in the socket
- [x] B: switch it off and remove the plug from the socket
- [ ] C: do nothing in particular
- [ ] D: put tape over the ON/OFF switch

10.4

Someone near you is using a disc cutter to cut concrete blocks. What three immediate hazards are likely to affect you?

- [x] A: Flying fragments
- [ ] B: Dermatitis
- [x] C: Dust in the air
- [x] D: High noise levels
- [ ] E: Skin cancer

10.5

You need to use a power tool to cut or grind materials. Give two ways to control the dust.

- [ ] A: Work slowly and carefully
- [x] B: Fit a dust extractor or collector to the machine
- [x] C: Wet cutting
- [ ] D: Keep the area clean and tidy
- [ ] E: Wear a dust mask or respirator

4/5

Answers: 10.1 = C, 10.2 = D, 10.3 = B, 10.4 = A,C,D, 10.5 = B,C

10.6

If you use a power tool to cut or grind materials, why must the dust be collected and not get into the air?

- A: To save time and avoid having to clear up the mess
- B: Most dust can be harmful if breathed in
- C: The tool will go faster if the dust is collected
- D: You do not need a machine guard if the dust is collected

10.7

If the head on your hammer comes loose, you should:

- A: stop work and get the hammer repaired or replaced
- B: find another heavy tool to use instead of the hammer
- C: keep using it but be aware that the head could come off at any time
- D: tell the other people near you to keep out of the way

10.8

If the guard is missing from a power tool, you should:

- A: try to make another guard
- B: use the tool but try to work quickly
- C: not use the tool until a proper guard has been fitted
- D: use the tool but work carefully and slowly

10.9

You must be fully trained before you use a cartridge-operated tool. Why?

- A: They are heavy and could cause manual handling injuries
- B: They operate like a gun and can be dangerous in inexperienced hands
- C: Using one can cause dermatitis
- D: They have exposed electrical parts

3/4

Answers: 10.6 = B, 10.7 = A, 10.8 = C, 10.9 = B

10.10

If you need to use a power tool with a rotating blade, you should:

- A: remove the guard so that you can clearly see the blade
- B: adjust the guard to expose just enough blade to let you do the job
- C: remove the guard but wear leather gloves to protect your hands
- D: adjust the guard to expose the maximum amount of blade

10.11

Chainsaws are dangerous because:

- A: they are heavy to use
- B: they are noisy
- C: there is no guard on the cutting chain
- D: all of the other answers

10.12

Before you use a power tool, you should check:

- A: it is not damaged and is fit to use
- B: it has your company's name and logo on it
- C: the serial number is clearly shown
- D: all of the other answers

10.13

If you need to use a grinding tool, what type of eye protection will you need?

- A: High impact eye protection
- B: Welding goggles
- C: None
- D: Prescription safety glasses or sunglasses

10.14

Most cutting and grinding machines have guards. What are the two main functions of the guard?

- A: To stop materials getting onto the blade or wheel
- B: To give you a firm handhold
- C: To balance the machine
- D: To stop fragments flying into the air
- E: To stop you coming into contact with the blade or wheel

Answers: 10.10 = B, 10.11 = D, 10.12 = A, 10.13 = A, 10.14 = D,E

10.15

You need to use a power tool, but it has a 13 amp plug fitted. What would be the safest thing to do?

- A: Fit a 110 volt plug so that you can plug it into the transformer
- B: Get a 110 volt tool to use
- C: Look for a 110/230 volt step-up transformer
- D: Find a suitable 230 volt supply to power it

10.16

Do you need to inspect simple hand tools like trowels, chisels and hammers?

- A: No, never
- B: Yes, if they have not been used for a couple of weeks
- C: Yes, they should be checked each time you use them
- D: Only if someone else has borrowed them

10.17

Someone near you is using a laser level. What health hazard is likely to affect you?

- A: Skin cancer
- B: None if it is used correctly
- C: Gradual blindness
- D: Burning of the skin, similar to sunburn

10.18

What is the main danger if you use a chisel with a 'mushroomed' head?

- A: It will shatter and send fragments flying into the air
- B: It will damage the face of the hammer
- C: The shaft of the chisel will bend when you hit it
- D: You will have to sharpen the chisel more often

10.19

Look at these statements about power tools. Which one is true?

- A: Always carry the tool by its cord
- B: Always unplug the tool by pulling its cord
- C: Always unplug the tool when you are not using it
- D: Always leave the tool plugged in when you check or adjust it

Answers: 10.15 = B, 10.16 = C, 10.17 = B, 10.18 = A, 10.19 = C

10.20

It is dangerous to run an abrasive wheel faster than its recommended top speed. Why?

- [ ] A: The wheel will get clogged and stop
- [ ] B: The motor could burst into flames
- [ ] C: The wheel could burst
- [ ] D: The safety guard cannot be used

10.21

You need to use an air-powered tool. Which of these is not a hazard?

- [ ] A: Electric shock
- [ ] B: Hard-arm vibration
- [ ] C: Airborne dust and flying fragments
- [ ] D: Leaking hoses

Answers: 10.20 = C, 10.21 = A

11.1

How does this type of extinguisher put out fires?

- [ ] A: It gets rid of the heat
- [ ] B: It keeps out oxygen
- [ ] C: It removes the fuel
- [ ] D: It makes the fire wet

11.2

This fire extinguisher contains:

- [ ] A: foam
- [ ] B: carbon dioxide (CO^2)
- [ ] C: water
- [ ] D: dry powder

11.3

You need to work in a corridor that is a fire escape route. You must see that:

- [ ] A: your tools and equipment do not block the route
- [ ] B: all doors into the corridor are locked
- [ ] C: you only use spark-proof tools
- [ ] D: you remove all fire escape signs before you start

11.4

A fire assembly point is the place where:

- [ ] A: fire engines must go when they arrive on site
- [ ] B: the fire extinguishers are kept
- [ ] C: people must go when the fire alarm sounds
- [ ] D: the fire started

11.5

To put out an oil fire, you must **not** use:

- [ ] A:
- [ ] B:
- [ ] C:
- [ ] D:

11.6

This fire extinguisher contains:

- [ ] A: water
- [ ] B: foam
- [ ] C: dry powder
- [ ] D: carbon dioxide (CO^2)

Answers: 11.1 = B, 11.2 = C, 11.3 = A, 11.4 = C, 11.5 = D, 11.6 = D

11.7

When you use a carbon dioxide (CO^2) fire extinguisher, the nozzle gets:

- A: very cold
- B: very hot
- C: warm
- D: very heavy

11.8

A hot work permit lets you:

- A: work in hot weather
- B: carry out work that needs warm protective clothing
- C: carry out work that could start a fire
- D: light a bonfire

11.9

If your job needs a hot work permit, what two things would you expect to have to do?

- A: Have a fire extinguisher close to the work
- B: Check for signs of fire when you stop work
- C: Know where all the fire extinguishers are kept on site
- D: Write a site evacuation plan
- E: Know how to refill fire extinguishers

11.10

Look at these jobs. Which two are likely to need a hot work permit?

- A: Cutting steel with an angle grinder
- B: Soldering pipework in a central heating system
- C: Refuelling a diesel dump truck
- D: Replacing an empty Liquid Petroleum Gas cylinder with a full one
- E: Working at night using halogen floodlights

11.11

All fires need heat, fuel and:

- A: oxygen
- B: carbon dioxide
- C: argon
- D: nitrogen

Answers: 11.7 = A, 11.8 = C, 11.9 = A,B, 11.10 = A,B, 11.11 = A

11.12

If you discover a large fire, the first thing you should do is:

- [ ] A: put your tools away
- [ ] B: finish what you are doing if it is safe to do so
- [ ] C: try to put out the fire
- [x] D: raise the alarm

11.13

If you hear the fire alarm, you should go to:

- [ ] A: the site canteen
- [x] B: the assembly point
- [ ] C: the site office
- [ ] D: the fire

11.14

This extinguisher can be used to put out:

- [ ] A: burning oil
- [ ] B: electrical fires
- [x] C: wood fires
- [ ] D: burning petrol

6/6

11.15

A large fire has been reported. You have not been trained to use fire extinguishers. You should:

- [ ] A: put away all your tools and then go to assembly point
- [ ] B: report to the site office and then go home
- [x] C: go straight to the assembly point
- [ ] D: leave work for the day

11.16

This extinguisher must not be used on:

- [x] A: electrical fires
- [ ] B: wood fires
- [ ] C: burning furniture
- [ ] D: burning clothes

11.17

This extinguisher must not be used on:

- [x] A: electrical fires
- [ ] B: wood fires
- [ ] C: burning oil
- [ ] D: burning petrol

Answers: 11.12 = D, 11.13 = B, 11.14 = C, 11.15 = C, 11.16 = A, 11.17 = A

11.18

If you see 'frost' around the valve on a Liquid Petroleum Gas cylinder, it means:

- [ ] A: the cylinder is nearly empty
- [ ] B: the cylinder is full
- [x] C: the valve is leaking
- [ ] D: you must lay the cylinder on its side

11.19

If there is a fire, you will need to go to the site assembly point. When would you expect to be told where this is?

- [ ] A: By the Health and Safety Executive
- [x] B: During site induction
- [ ] C: By reading your employer's health and safety policy
- [ ] D: Your mates will tell you

11.20

Which two extinguishers can be used on electrical fires?

- [x] A:
- [x] B:
- [ ] C:
- [ ] D:

11.21

Which two extinguishers are best for putting out oil fires?

- [ ] A:
- [x] B:
- [x] C:
- [ ] D:

11.22

This type of fire extinguisher puts out a fire by:

- [ ] A: starving the fire of oxygen
- [x] B: cooling the burning material
- [ ] C: increasing the oxygen supply
- [ ] D: diluting the oxygen with another gas

3/5

Answers: 11.18 = C, 11.19 = B, 11.20 = A,D, 11.21 = A,C, 11.22 = B

12.1

Which sign means flammable substance?

- ☐ A:
- ☐ B:
- ☐ C:
- ☐ D:

12.2

This sign means:

- ☐ A: wear ear protection if you want to
- ☐ B: you must wear ear protection
- ☐ C: you must not make a noise
- ☐ D: caution, noisy machinery

12.3

Fire exit signs are coloured:

- ☐ A: blue and white
- ☐ B: red and white
- ☐ C: green and white
- ☐ D: red and yellow

12.4

This sign tells you where:

- ☐ A: to go if there is a fire
- ☐ B: fire extinguishers should be kept
- ☐ C: a fire will start
- ☐ D: flammable materials should be kept

12.5

This sign means:

- ☐ A: press here to sound the fire alarm
- ☐ B: do not touch
- ☐ C: wear hand protection
- ☐ D: press here to switch on the emergency light

Answers: 12.1 = B, 12.2 = B, 12.3 = C, 12.4 = B, 12.5 = A

Safety Signs and Signals

12.6

This sign means:

- A: press here to sound the fire alarm
- B: fire hose reel
- C: turn key to open
- D: do not use if there is a fire

12.7

This sign means:

- A: danger from radiation
- B: danger from bright lights or lasers
- C: caution, poor lighting
- D: you must wear eye protection

12.8

Which sign means 'Warning, laser beams'?

- A:
- B:
- C:
- D:

12.9

This sign means:

- A: smoking allowed
- B: danger, flammable materials
- C: no smoking
- D: no explosives

12.10

This sign means:

- A: no access onto the scaffold
- B: no entry without full Personal Protective Equipment
- C: no entry for people on foot
- D: no entry during the day

12.11

Green and white signs mean:

- A: you **must** do something
- B: you **must not** do something
- C: hazard or danger
- D: safe condition

6/6

Answers: 12.6 = B, 12.7 = D, 12.8 = D, 12.9 = C, 12.10 = C, 12.11 = D

12.12

Blue and white signs mean:

- [x] A: you **must** do something
- [ ] B: you **must not** do something
- [ ] C: hazard or danger
- [ ] D: safe condition

12.13

Yellow and black signs mean:

- [ ] A: you **must** do something
- [ ] B: you **must not** do something
- [x] C: hazard or danger
- [ ] D: safe condition

12.14

Red and white signs with a red line mean:

- [ ] A: you **must** do something
- [x] B: you **must not** do something
- [ ] C: hazard or danger
- [ ] D: safe condition

12.15

This sign means:

- [ ] A: plant operators wanted
- [x] B: forklift trucks operating
- [ ] C: manual handling not allowed
- [ ] D: storage area

12.16

If you see this sign on a scaffold, you should:

- [ ] A: remove the access ladder
- [ ] B: only work on the first lift
- [x] C: stay off the scaffold because it is not safe
- [ ] D: only use a Mobile Elevating Work Platform to get on to the scaffold

12.17

This sign means:

- [x] A: you must wear safety boots
- [ ] B: you must wear Wellington boots
- [ ] C: caution, slip and trip hazards
- [ ] D: wear safety boots if you want to

6/6

Answers: 12.12 = A, 12.13 = C, 12.14 = B, 12.15 = B, 12.16 = C, 12.17 = A

12.18

This sign means:

- [ ] A: caution, cold materials
- [ ] B: caution, hot materials
- [ ] C: carry out work using one hand
- [x] D: you must wear safety gloves

12.19

If you see this sign, you must:

- [ ] A: wear white clothes at night
- [x] B: wear high visibility clothes
- [ ] C: do nothing, it only applies to managers
- [ ] D: wear wet weather clothes

5/5

12.20

A crane has to do a difficult lift. The signaller asks you to help, but you are not trained in plant signals. What should you do?

- [x] A: Politely refuse because you don't know how to signal
- [ ] B: Start giving signals to the crane driver
- [ ] C: Only help if the signaller really can't manage alone
- [ ] D: Ask the signaller to show you what signals to use

12.21

A truck has to tip materials into a trench. Who should give signals to the truck driver?

- [ ] A: Anyone who is there
- [ ] B: Someone standing in the trench
- [ ] C: Anyone who knows the signals
- [x] D: Anyone who is trained and competent

12.22

This sign means:

- [ ] A: leaking roof
- [ ] B: wear waterproof clothes
- [x] C: emergency shower
- [ ] D: fire sprinklers

Answers: 12.18 = D, 12.19 = B, 12.20 = A, 12.21 = D, 12.22 = C

12.23

This sign means:

- [ ] A: do not run
- [ ] B: no escape route
- [ ] C: fire door
- [ ] D: fire escape route

12.24

This sign tells you that a substance is:

- [ ] A: harmful
- [ ] B: toxic
- [ ] C: corrosive
- [ ] D: dangerous to the environment

Answers: 12.23 = B, 12.24 = D

Site Transport Safety

13.1

What are the two conditions for being able to operate plant on site?

- [ ] A: You must be competent
- [x] B: You must be authorised
- [ ] C: You must be over 21 years old
- [x] D: You must hold a full driving licence
- [ ] E: You must hold a UK passport

13.2

A mobile plant operator can let you ride in the machine:

- [ ] A: if you have a long way to go
- [ ] B: if it is raining
- [x] C: if it is designed to carry passengers
- [ ] D: at any time

13.3

You think some mobile plant is operating too close to where you have to work. What should you do?

- [ ] A: Stop work and speak to the plant operator
- [ ] B: Stop work and speak to the plant operator's supervisor
- [ ] C: Keep a good look-out for the plant and carry on working
- [x] D: Stop work and speak to your own supervisor

13.4

You need to walk past someone using a mobile crane. You should:

- [ ] A: guess what the crane operator will do next and squeeze by
- [ ] B: try to catch the attention of the crane operator
- [ ] C: run to get past the crane quickly
- [x] D: take another route so that you stay clear of the crane

13.5

When you walk across the site, what is the best way to avoid an accident with mobile plant?

- [x] A: Keep to the pedestrian routes
- [ ] B: Ride on the plant
- [ ] C: Get the attention of the driver before you get too close
- [ ] D: Wear high visibility clothing

4/5

Answers: 13.1 = A,B, 13.2 = C, 13.3 = D, 13.4 = D, 13.5 = A

13.6

You need to walk past a 360° mobile crane. The crane is operating near a wall. What is the main danger?

- [ ] A: The crane could crash into the wall
- [x] B: You could be crushed if you walk between the crane and the wall
- [ ] C: Whole-body vibration from the crane
- [ ] D: High noise levels from the crane

13.7

Your supervisor asks you to drive a dumper truck. You have never driven one before. What should you do?

- [ ] A: Ask a trained driver how to operate it
- [x] B: Tell your supervisor that you cannot operate it
- [ ] C: Watch other dumpers to see how they are operated
- [ ] D: Get on with it

5/5

13.8

You are walking across the site. A large mobile crane reverses across your path. What should you do?

- [ ] A: Help the driver to reverse
- [ ] B: Start to run so that you can pass behind the reversing crane
- [ ] C: Pass close to the front of the crane
- [x] D: Wait or find another way around the crane

13.9

If you see a dumper being driven too fast, you should:

- [x] A: keep out of its way and report the matter
- [ ] B: try to catch the dumper and speak to the driver
- [ ] C: report the matter to the police
- [ ] D: do nothing, dumpers are allowed to go above the site speed limit

13.10

When is site transport allowed to drive along a pedestrian route?

- [ ] A: During meal breaks
- [ ] B: If it is the shortest route
- [x] C: Only if necessary and if all pedestrians are excluded
- [ ] D: Only if the vehicle has a flashing yellow light

Answers: 13.6 = B, 13.7 = B, 13.8 = D, 13.9 = A, 13.10 = C

13.11

Which of these would you not expect to see if site transport is well organised?

- [ ] A: Speed limits
- [ ] B: Barriers to keep pedestrians away from mobile plant and vehicles
- [x] C: Pedestrians and mobile plant using the same routes
- [ ] D: One-way systems

13.12

A lorry is in trouble as it tries to reverse into a tight space. You have not been trained as a signaller. What should you do?

- [x] A: Stay well out of the way
- [ ] B: Help the driver by giving hand signals
- [ ] C: Help the driver by jumping up into the cab
- [ ] D: Offer to adjust the mirrors on the lorry

13.13

You see a lorry parking. It has a flat tyre. Why should you tell the driver?

- [ ] A: The lorry will use more fuel
- [ ] B: The lorry will need to travel at a much slower speed
- [x] C: The lorry is unsafe to drive
- [ ] D: The lorry can only carry small loads

13.14

An excavator has just stopped work. Liquid is dripping and forming a small pool under the back of the machine. What could this mean?

- [ ] A: It is normal for fluids to vent after the machine stops
- [ ] B: The machine is hot so the diesel has expanded and overflowed
- [ ] C: Someone put too much diesel into the machine before it started work
- [x] D: The machine has a leak and could be unsafe

13.15

You see a mobile crane lifting a load. The load is about to hit something. What should you do?

- [ ] A: Go and tell your supervisor
- [x] B: Tell the person supervising the lift
- [ ] C: Go and tell the crane driver
- [ ] D: Do nothing and assume everything is under control

5/5

Answers: 13.11 = C, 13.12 = A, 13.13 = C, 13.14 = D, 13.15 = B

13.16

You see a driver refuelling an excavator. Most of the diesel is spilling on to the ground. What is the first thing you should do?

- [ ] A: Tell your supervisor the next time you see them
- [x] B: Tell the driver immediately
- [ ] C: Look for a spillage kit immediately
- [ ] D: Do nothing, the diesel will eventually seep into the ground

13.17

How would you expect a well-organised site to keep pedestrians away from traffic routes?

- [ ] A: The site manager will direct all pedestrians away from traffic routes
- [ ] B: The traffic routes will be shown on the Health and Safety Law poster
- [x] C: There will be barriers between traffic and pedestrian routes
- [ ] D: There is no need to keep traffic and pedestrians apart

13.18

A site vehicle is most likely to injure pedestrians when it is

- [x] A: reversing
- [ ] B: lifting materials onto scaffolds
- [ ] C: tipping into an excavation
- [ ] D: digging out footings

13.19

You must not walk behind a lorry when it is reversing because:

- [ ] A: most lorries are not fitted with mirrors
- [x] B: the driver is unlikely to know you are there
- [ ] C: most lorry drivers aren't very good at reversing
- [ ] D: you will need to run, not walk, to get past it in time

13.20

The quickest way to your work area is through a contractor's vehicle compound. Which way should you go?

- [ ] A: Around the compound if vehicles are moving
- [ ] B: Straight through the compound if no vehicles appear to be moving
- [x] C: Around the compound every time
- [ ] D: Straight through the compound if no-one is looking

Answers: 13.16 = B, 13.17 = C, 13.18 = A, 13.19 = B, 13.20 = C

13.21

How would you expect to be told about the site traffic rules?

- [x] A: During site induction
- [ ] B: By an Health and Safety Executive inspector
- [ ] C: By a note on a notice board
- [ ] D: In a letter sent to your home

13.22

A forklift truck is blocking the way to your work area. It is lifting materials on to a scaffold. What should you do?

- [ ] A: Only walk under the raised load if you are wearing a safety helmet
- [ ] B: Catch the driver's attention and then walk under the raised load
- [ ] C: Start to run so that you are not under the load for very long
- [x] D: Wait or go around, but never walk under a raised load

13.23

What do you need before you can supervise a lift using a crane?

- [ ] A: Nothing, you make it up as you go along
- [x] B: You must be trained and assessed as competent
- [ ] C Written instructions from the crane hire company
- [ ] D: Nothing, the crane driver will tell you what to do

13.24

You think a load is about to fall from a moving forklift truck. What should you do?

- [x] A: Keep clear but try to warn the driver and others in the area
- [ ] B: Run alongside the machine and try to hold on to the load
- [ ] C: Run and tell your supervisor
- [ ] D: Sound the nearest fire alarm bell

Answers: 13.21 = A, 13.22 = D, 13.23 = B, 13.24 = A

14.1

Noise can damage your hearing. What is an early sign of this?

- [x] A: There are no early signs
- [ ] B: Temporary deafness
- [ ] C: A skin rash around the ears
- [ ] D: Ear infections

14.2

After working with noisy equipment, you have a 'ringing' sound in your ears. What does this mean?

- [x] A: Your hearing has been temporarily damaged
- [ ] B: You have also been subjected to vibration
- [ ] C: You are about to go down with the 'flu
- [ ] D: The noise level was high but acceptable

14.3

If you wear hearing protection, it will:

- [ ] A: stop you hearing all noise
- [x] B: reduce noise to an acceptable level
- [ ] C: repair your hearing if it is damaged
- [ ] D: make you hear better

14.4

Noise over a long time can damage your hearing. Can this damage be reversed?

- [ ] A: Yes, with time
- [ ] B: Yes, if you have an operation
- [x] C: No, the damage is permanent
- [ ] D: Yes, if you change jobs

14.5

If you need to wear hearing protection, you must remember that:

- [ ] A: you have to carry out your own noise assessment
- [ ] B: you have to pay for all hearing protection
- [ ] C: ear plugs don't work
- [x] D: you may be less aware of what is going on around you

14.6

Two recommended ways to protect your hearing are:

- [ ] A: rolled tissue paper
- [ ] B: cotton wool pads
- [x] C: ear plugs
- [ ] D: soft cloth pads
- [x] E: ear defenders

Answers: 14.1 = B, 14.2 = A, 14.3 = B, 14.4 = C, 14.5 = D, 14.6 = C,E

14.7

How can noise affect your health? Give two answers.

- [x] A: Headaches
- [ ] B: Ear infections
- [x] C: Hearing loss
- [ ] D: Waxy ears
- [ ] E: Vibration white finger

14.8

Noise may be a problem in your work area if you have to shout to be clearly heard by someone who is standing:

- [ ] A: 2 Metres away
- [x] B: 4 Metres away
- [ ] C: 5 Metres away
- [ ] D: 6 Metres away

14.9

You think the noise at work may have damaged your hearing. What should you do?

- [ ] A: Plug your ears with cotton wool to stop any more damage
- [ ] B: Nothing, the damage has already been done
- [ ] C: Go off sick
- [x] D: Ask your employer or doctor to arrange a hearing test

5/6.

14.10

You need to wear ear defenders, but an ear pad is missing from one of the shells. What should you do?

- [ ] A: Leave them off and work without any hearing protection
- [ ] B: Put them on and start working with them as they are
- [x] C: Do not work in noisy areas until they are replaced
- [ ] D: Wrap your handkerchief around the shell and carry on working

14.11

If you have to work in an 'ear protection zone', you must:

- [ ] A: not make any noise
- [x] B: wear hearing protection at all times
- [ ] C: take hearing protection with you in case you need to use it
- [ ] D: wear hearing protection if the noise gets too loud for you

14.12

Why is vibration a serious health issue?

- [ ] A: There are no early warning signs
- [ ] B: The long-term effects of vibration are not known
- [ ] C: There is no way that exposure to vibration can be prevented
- [x] D: Vibration can cause a disabling injury that cannot be cured

Answers: 14.7 = A,C, 14.8 = A, 14.9 = D, 14.10 = C, 14.11 = B, 14.12 = D

14.13

What are three early signs of vibration white finger?

- [x] A: Temporary loss of feeling in the fingers
- [x] B: The fingertips turn white
- [ ] C: A skin rash
- [x] D: Tingling in the fingers
- [ ] E: Blisters

14.14

You are less likely to suffer from hand-arm vibration if you are:

- [ ] A: very cold but dry
- [ ] B: cold and wet
- [x] C: warm and dry
- [ ] D: very wet but warm

14.15

If you need to use a vibrating tool, how can you help reduce the risk of hand-arm vibration?

- [x] A: Do not grip the tool too tightly
- [ ] B: Hold the tool away from you, at arms' length
- [ ] C: Use more force
- [ ] D: Hold the tool more tightly

14.16

Hand-arm vibration can cause:

- [ ] A: skin cancer
- [ ] B: skin irritation, like dermatitis
- [ ] C: blisters on your hands and arms
- [x] D: damaged blood vessels and nerves in your fingers and hands

14.17

If you have to use a vibrating tool, what would you expect your supervisor to do?

- [ ] A: Measure the level of vibration while you use the tool
- [x] B: Tell you about the risk assessment and explain the safe way to use the tool
- [ ] C: Watch you use the tool to assess the level of vibration
- [ ] D: Help you to make up your own safe system of work

5/5

Answers: 14.13 = A,B,D, 14.14 = C, 14.15 = A, 14.16 = D, 14.17 = B

Noise and Vibration

14.18

You have been using a vibrating tool. The end of your fingers are starting to tingle. What does this mean?

- A: You can carry on using the tool but you must loosen your grip
- B: You must not use this tool, or any other vibrating tool, ever again
- C: You need to report your symptoms before they cause a problem
- D: You can carry on using the tool but you must hold it tighter

14.19

What is vibration white finger?

- A: A mild skin rash that will go away
- B: A serious skin condition that will not clear up
- C: Severe frostbite
- D: A sign of damage to your hands and arms that might not go away

14.20

If you have to use a vibrating tool, how can you help reduce the effects of hand-arm vibration?

- A: Hold the tool tightly
- B: Do the work in short spells
- C: Do the job in one long burst
- D: Only use one hand on the tool at a time

14.21

Which of these is most likely to cause vibration white finger?

- A: Electric hoist
- B: Hammer drill
- C: Hammer and chisel
- D: Battery-powered screwdriver

14.22

Someone near you is using noisy equipment and you have no hearing protection. What should you do?

- A: Ask them to stop what they are doing
- B: Carry on with your work because it is always noisy on site
- C: Leave the area until you have the correct PPE
- D: Speak to the other person's supervisor

Answers: 14.18 = C, 14.19 = D, 14.20 = B, 14.21 = B, 14.22 = C

15.1

What must happen each time a shift starts work in an excavation?

- [ ] A: Someone must go in and sniff the air to see if it is safe
- [x] B: A competent person must inspect the excavation
- [ ] C: A supervisor should stay in the excavation for the first hour
- [ ] D: A supervisor should watch from the top for the first hour

15.2

When digging, you hit and damage a buried cable. What should you do?

- [ ] A: Move the cable out of the way and carry on digging
- [ ] B: Wait 10 seconds and then move the cable out of the way
- [x] C: Do not touch the cable, stop work and report it
- [ ] D: Dig round the cable or dig somewhere else

15.3

An excavation must be supported if:

- [ ] A: it is more than 5 metres deep
- [x] B: it is more than 1.2 metres deep
- [ ] C: there is a risk of the sides falling in
- [ ] D: any buried services cross the excavation

15.4

You are working in an excavation. If you see the side supports move, you should:

- [ ] A: keep watching to see if they move again
- [x] B: make sure that you and others get out quickly
- [ ] C: do nothing as the sides move all the time
- [ ] D: work in another part of the excavation

15.5

What is the safe way to get into a deep excavation?

- [x] A: Climb down a ladder
- [ ] B: Use the buried services as steps
- [ ] C: Climb down the shoring
- [ ] D: Go down in an excavator bucket

Answers: 15.1 = B, 15.2 = C, 15.3 = C, 15.4 = B, 15.5 = A

15.6

When digging, you find a run of yellow plastic marker tape. What does it mean?

- [ ] A: There are buried human remains and you must tell your supervisor
- [x] B: There is a buried service and further excavation must be carried out with care
- [ ] C: The soil is contaminated and you must wear Respiratory Protective Equipment
- [ ] D: The excavation now needs side supports

15.7

Which of these is the most accurate way to locate buried services?

- [ ] A: Cable plans
- [x] B: Trial holes
- [ ] C: Survey drawings
- [ ] D: Architect's drawings

15.8

If you need to dig near underground services, you should use:

- [ ] A: a jack hammer
- [x] B: a spade or shovel
- [ ] C: a pick and fork
- [ ] D: an excavator

15.9

You are in a deep trench. A lorry backs up to the trench and the engine is left running. What should you do?

- [ ] A: Put on ear defenders to cut out the engine noise
- [ ] B: Ignore the problem, the lorry will soon drive away
- [ ] C: See if there is a toxic gas meter in the trench
- [x] D: Get out of the trench quickly

15.10

You are in a deep trench and start to feel dizzy. What should you do?

- [ ] A: Get out, let your head clear and then go back in again
- [ ] B: Carry on working and hope that the feeling will go away
- [x] C: Make sure that you and any others get out quickly
- [ ] D: Sit down in the trench and take a rest

5/5.

Answers: 15.6 = B, 15.7 = B, 15.8 = B, 15.9 = D, 15.10 = C

15.11

Before work starts in a confined space, how should the air be checked?

- [ ] A: Someone should go in and sniff the air
- [x] B: The air should be tested with a meter
- [ ] C: Someone should look around to see if there is toxic gas
- [ ] D: The air should be tested with a match to see if it stays alight

15.12

If you need to work in a confined space, one duty of the top man is to:

- [ ] A: tell you how to work safely in confined spaces
- [ ] B: enter the confined space if there is a problem
- [x] C: start the rescue plan if needed
- [ ] D: supervise the work in the confined space

15.13

Which of these is not a hazard in a confined space?

- [ ] A: Toxic gas
- [x] B: A lack of carbon dioxide
- [x] C: A lack of oxygen
- [ ] D: Flammable or explosive gas

6/6

15.14

If there is sludge at the bottom of a confined space, you should:

- [ ] A: go in and then step into the sludge to see how deep it is
- [ ] B: throw something into the sludge to see how deep it is
- [ ] C: put on a disposable face-mask before you go in
- [x] D: have the correct Respiratory Protective Equipment and training before you go in

15.15

Why is methane gas dangerous in confined space? Give two answers.

- [x] A: It can explode
- [ ] B: It makes you hyperactive
- [ ] C: You will not be able to see because of the dense fumes
- [ ] D: It makes you dehydrated
- [x] E: You will not have enough oxygen to breathe

15.16

You are in a confined space. If the level of oxygen drops:

- [ ] A: your hearing could be affected
- [ ] B: there is a high risk of fire or explosion
- [x] C: you could become unconscious
- [ ] D: you might get dehydrated

Answers: 15.11 = B, 15.12 = C, 15.13 = B, 15.14 = D, 15.15 = A,E, 15.16 = C

Excavations and Confined Spaces

15.17

You have to work in a confined space. There is no rescue team or rescue plan. What should you do?

- [ ] A: Assume that a rescue team or plan is not necessary and do the job
- [ ] B: Get someone to stand at the opening with a rope
- [x] C: Do not enter until a rescue plan and team are in place
- [ ] D: Carry out the job in short spells

15.18

You are working in a confined space when you notice the smell of bad eggs. This smell is a sign of:

- [x] A: hydrogen sulphide
- [ ] B: oxygen
- [ ] C: methane
- [ ] D: carbon dioxide

15.19

You need to walk through sludge at the bottom of a confined space. Which of these is not a hazard?

- [x] A: The release of oxygen
- [ ] B: The release of toxic gases
- [ ] C: Slips and trips
- [ ] D: The release of flammable gases

15.20

You are working in a confined space. If the Permit to Work runs out before you finish the job, you should:

- [ ] A: carry on working until the job is finished
- [ ] B: hand the permit over to the next shift
- [ ] C: ask your supervisor to change the date on the permit
- [x] D: leave the confined space before the permit runs out

15.21

When digging, you notice the soil gives off a strange smell. What is this likely to mean?

- [ ] A: The soil contains a lot of clay
- [ ] B: The soil has been excavated before
- [ ] C: The ground has been used to grow crops in the past
- [x] D: The ground could be contaminated

5/5

Answers: 15.17 = C, 15.18 = A, 15.19 = A, 15.20 = D, 15.21 = D

15.22

Guard-rails are placed around the top of an excavation to prevent:

- [ ] A: toxic gases from collecting in the bottom of the trench
- [x] B: anyone falling into the trench and being injured
- [ ] C: the sides of the trench from collapsing
- [ ] D: rain water running off the ground at the top and into the trench

15.23

Work in a confined space usually needs three safety documents - a risk assessment, a method statement and:

- [x] A: a Permit to Work
- [ ] B: an up-to-date staff handbook
- [ ] C: a written contract for the work
- [ ] D: a company health and safety policy

15.24

You are in a confined space when the gas alarm sounds. You have no Respiratory Protective Equipment, what should you do?

- [ ] A: Switch off the alarm
- [x] B: Get out of the confined space quickly
- [ ] C: Carry on working but do not use electrical tools
- [ ] D: Carry on working but take plenty of breaks in the fresh air

3/3

Answers: 15.22 = B, 15.23 = A, 15.24 = B

If you are taking one of the specialist tests, you will be asked questions from Sections 1 - 15 plus questions on one of the following subjects:

- Section 16 Supervisory and Management
- Section 17 Demolition
- Section 18 Plumbing
- Section 19 Highway Works
- Section 20 Specialist Working at Height
- Section 21 Lifts and Escalators

Supervisory and Management

16.1

What is the purpose of the health and safety file on a construction project?

- [ ] A: To assist people who have to carry out work on the structure in the future
- [ ] B: To assist in the preparation of final accounts for the structure
- [ ] C: To record the health and safety standards of the structure
- [ ] D: To record the accident details

16.2

Under the Construction (Design and Management) Regulations, what must be in place before construction work begins?

- [ ] A: Construction project health and safety file
- [ ] B: Construction phase health and safety plan
- [ ] C: Construction project plan
- [ ] D: Construction contract agreement

16.3

Which form must be displayed on projects where the Construction (Design and Management) Regulations apply?

- [ ] A: Form F9
- [ ] B: Form F10 (Rev)
- [ ] C: Form 11
- [ ] D: Form 12

16.4

Under the Construction (Design and Management) Regulations, who is responsible for ensuring notification to the Health and Safety Executive of the project?

- [ ] A: Client
- [ ] B: Designer
- [ ] C: Planning supervisor
- [ ] D: Principal contractor

16.5

If the Construction (Design and Management) Regulations apply on a project, construction must not start until which one of the following criteria has been met?

- [ ] A: The construction phase health and safety plan is in place
- [ ] B: A site manager has been employed to take charge
- [ ] C: The Health and Safety Executive has given permission
- [ ] D: The health and safety file is in place

Answers: 16.1 = A, 16.2 = B, 16.3 = B, 16.4 = C, 16.5 = A

16.6

The planning supervisor has responsibilities for which two activities under the Construction (Design and Management) Regulations?

- [ ] A: Ensuring co-ordination of the health and safety aspects of the design
- [ ] B: Appointing the designer
- [ ] C: Deciding which construction processes are to be used
- [ ] D: Ensuring that the pre-tender health and safety plan is prepared
- [ ] E: Monitoring site safety

16.7

To whom should the planning supervisor pass the health and safety file on completion of the construction project?

- [ ] A: The Association of Planning Supervisors
- [ ] B: The client
- [ ] C: The Health and Safety Executive
- [ ] D: The designer

16.8

If, as a result of an accident at work, an employee is absent from work for more than three days, how soon must the Health and Safety Executive be notified?

- [ ] A: Within 3 days
- [ ] B: Within 5 days
- [ ] C: Within 7 days
- [ ] D: Within 10 days

16.9

If there is a fatal accident on site, when must the Health and Safety Executive be informed?

- [ ] A: Immediately
- [ ] B: Within 5 days
- [ ] C: Within 7 days
- [ ] D: Within 10 days

16.10

Following a reportable dangerous occurrence, when must the Health and Safety Executive be informed?

- [ ] A: Within 1 day
- [ ] B: Within 5 days
- [ ] C: Within 10 days
- [ ] D: Immediately

Answers: 16.6 = A,D, 16.7 = B, 16.8 = D, 16.9 = A, 16.10 = D

16.11

If an employee is injured in an accident that results in time off work, when must you report it to the Health and Safety Executive under the Reporting of Injuries, Diseases and Dangerous Occurrences Regulations?

- [ ] A: When over half a day is lost
- [ ] B: When over 1 day is lost
- [ ] C: When over 2 days are lost
- [ ] D: When over 3 days are lost

16.12

What is regarded as the last resort for operatives' safety when working at height?

- [ ] A: Safety harness
- [ ] B: Mobile elevating work platform
- [ ] C: Mobile tower scaffold
- [ ] D: Access tower scaffold

16.13

Which of the following is a fall-arrest system?

- [ ] A: Mobile access equipment
- [ ] B: Scaffold towers
- [ ] C: Mobile elevating work platform
- [ ] D: Safety harness and lanyard

16.14

Which colour identifies the 'live' wire in a 240 volt supply?

- [ ] A: Black
- [ ] B: Brown
- [ ] C: Green
- [ ] D: Yellow

16.15

Which piece of equipment is used with a cable avoidance tool (CAT) to detect cables?

- [ ] A: Compressor
- [ ] B: Signal generator
- [ ] C: Battery
- [ ] D: Gas detector

16.16

In the colour coding of electrical power supplies on site, what voltage does a blue plug represent?

- [ ] A: 50 volts
- [ ] B: 110 volts
- [ ] C: 240 volts
- [ ] D: 415 volts

Answers: 16.11 = D, 16.12 = A, 16.13 = D, 16.14 = B, 16.15 = B, 16.16 = C

16.17

On the site electrical distribution system, which colour plug indicates a 415 volt supply?

- [ ] A: Yellow
- [ ] B: Blue
- [ ] C: Black
- [ ] D: Red

16.18

When referring to noise, what does the term 'second action level' mean?

- [ ] A: Level at which hearing protection zones must be established and hearing protection must be worn
- [ ] B: Second time a noise reading is taken
- [ ] C: Time at which a second pair of ear defenders are provided
- [ ] D: When two noise meters are required

16.19

What does the term 'first action level' mean when referring to noise?

- [ ] A: Noise level at the start of the job
- [ ] B: Noise level at which the employee can request hearing protection
- [ ] C: Noise level when machines on the job first start up
- [ ] D: Noise level at which the employee must wear hearing protection

16.20

At what decibel (dBA) level does it become mandatory for an employer to establish hearing protection zones?

- [ ] A: 80 dBA
- [ ] B: 85 dBA
- [ ] C: 90 dBA
- [ ] D: 95 dBA

16.21

Which of the following is a significant hazard when excavating alongside a building or structure?

- [ ] A: Undermining the foundations of the building
- [ ] B: Upsetting the owners of the building
- [ ] C: Excavating too deep in soft ground
- [ ] D: Damage to the surface finish of the building or structure

Answers: 16.17 = D, 16.18 = A, 16.19 = B, 16.20 = C, 16.21 = A

16.22

What danger is created by excessive oxygen in a confined space?

- [ ] A: Increase in breathing rate of workers
- [ ] B: Increased flammability of combustible materials
- [ ] C: Increased working time inside work area
- [ ] D: False sense of security

16.23

When planning possible work in a confined space, what should be the first consideration?

- [ ] A: How long the job will take
- [ ] B: To avoid the need for operatives to enter the space
- [ ] C: How many operatives will be required
- [ ] D: Personal protective equipment

16.24

At what minimum noise level must you provide hearing protection to employees if they ask for it?

- [ ] A: 83 decibels
- [ ] B: 85 decibels
- [ ] C: 87 decibels
- [ ] D: 90 decibels

16.25

Employers must prevent exposure of their employees to substances hazardous to health. Where this is not reasonably practicable, which of the following should be done first?

- [ ] A: Provide instruction, training and supervision
- [ ] B: Carry out proper health surveillance
- [ ] C: Minimise risk and control exposure
- [ ] D: Monitor the exposure of employees in the workplace

16.26

The monitoring and controlling of health and safety procedures can be proactive or reactive. Reactive monitoring means:

- [ ] A: ensuring that staff are doing the work that they have been instructed to do
- [ ] B: looking at incidents after the event so that remedial action can be taken
- [ ] C: making sure that worksheets are up to date
- [ ] D: keeping a hazard book for use by all staff

Answers: 16.22 = B, 16.23 = B, 16.24 = B, 16.25 = C, 16.26 = B

16.27

Before planning for anyone to enter a confined space, what should be the first consideration of the manager or supervisor?

- [ ] A: Has the atmosphere in the confined space been tested?
- [ ] B: Has a safe means of access and egress been established?
- [ ] C: Is there an alternative method of doing the work?
- [ ] D: Have all operatives been properly trained?

16.28

What is the purpose of using a 'permit-to-work' system?

- [ ] A: To ensure that the job is being carried out properly
- [ ] B: To ensure that the job is carried out by the quickest method
- [ ] C: To enable tools and equipment to be properly checked before work starts
- [ ] D: To establish a safe system of work

16.29

In deciding which control measures to take, following a risk assessment which has revealed a risk, what measure should you always consider first?

- [ ] A: Make sure personal protective equipment is available
- [ ] B: Adapt the work to the individual
- [ ] C: Give priority to those measures which protect the whole workforce
- [ ] D: Avoid the risk altogether if possible

16.30

The monitoring and controlling of health and safety procedures can be either proactive or reactive. Proactive monitoring means:

- [ ] A: ensuring that staff always do the work that they have been instructed to do safely
- [ ] B: deciding how to prevent accidents similar to those that have already occurred
- [ ] C: looking at the work to be done, what could go wrong and how it could be done safely
- [ ] D: checking that all staff read and understand all health and safety notices

Answers: 16.27 = C, 16.28 = D, 16.29 = D, 16.30 = C

16.31

In considering what measures to take to protect people against risks to their health and safety, personal protective equipment should always be regarded as:

- [ ] A: the first line of defence
- [ ] B: the only practical measure
- [ ] C: the best way to tackle the job
- [ ] D: the last resort

16.32

Under the Construction (Design and Management) Regulations 1994, who is responsible for ensuring that the pre-tender health and safety plan is prepared?

- [ ] A: The planning supervisor
- [ ] B: The client
- [ ] C: A contractor tendering for the project
- [ ] D: The company safety officer

16.33

Under the Construction (Design and Management) Regulations, in which two of the following situations must the Health and Safety Executive be notified of a project?

- [ ] A: Where the work will last more than 30 days or more than 500 person-days
- [ ] B: Where the building and construction work will last more than 300 person-days
- [ ] C: Where the building and construction work is expected to last more than six months
- [ ] D: Where there is more than one building to be erected
- [ ] E: When the work will take place outside normal hours

16.34

An emergency route(s) must be provided on construction sites to ensure:

- [ ] A: safe passage to the ground
- [ ] B: safe passage to open air
- [ ] C: safe passage to a place of safety
- [ ] D: safe passage to the first-aid room

Answers: 16.31 = D, 16.32 = A, 16.33 = A,C, 16.34 = C

16.35

What is the best way for a supervisor or manager to make sure that the operatives doing a job have fully understood a method statement?

- [ ] A: Put the method statements in a labelled spring-binder in the office
- [ ] B: Explain the method statement to those doing the job
- [ ] C: Make sure that those doing the job have read the method statement
- [ ] D: Display the method statements on a notice board in the office

16.36

Under the Construction (Design and Management) Regulations, which duty-holder is responsible for assessing the competence and resources of the principal contractor?

- [ ] A: The planning supervisor
- [ ] B: The designer
- [ ] C: The client
- [ ] D: The safety officer

16.37

The Construction (Health, Safety and Welfare) Regulations 1996 require a supported excavation to be inspected:

- [ ] A: every 7 days
- [ ] B: at the start of every shift
- [ ] C: once a month
- [ ] D: when it is more than 2 metres deep

16.38

Before starting any construction work lasting more than 30 days, or 500 person-days, which of the following must be done?

- [ ] A: Inform the local authority on Form F9
- [ ] B: Locate all underground services
- [ ] C: Ensure all operatives are trained and competent
- [ ] D: Notify the Health and Safety Executive on Form F10 (Rev)

16.39

An employer has to prepare a written health and safety policy if:

- [ ] A: they employ 5 people or more
- [ ] B: they employ more than 3 people
- [ ] C: they employ a safety officer
- [ ] D: the work is going to last more than 30 days

Answers: 16.35 = B, 16.36 = C, 16.37 = B, 16.38 = D, 16.39 = A

16.40

Under the Work at Height Regulations, if part of a system-built scaffold is stopping someone from getting on with their job, the scaffold may be temporarily altered by:

- [ ] A: the person who is being prevented from working
- [ ] B: any site manager
- [ ] C: a trained and competent person
- [ ] D: anyone

16.41

The new-style Accident Book must now be used because it:

- [ ] A: contains the personal details of everyone on site
- [ ] B: can only be completed by a Site Manager or Supervisor
- [ ] C: complies with the requirements of the Data Protection Act
- [ ] D: can only be kept in an electronic format

16.42

When overhead electric cables cross a construction site, it is recommended that barriers should be erected parallel to the overhead cables at a distance not less than:

- [ ] A: 3 metres
- [ ] B: 4 metres
- [ ] C: 5 metres
- [ ] D: 6 metres

16.43

If a prohibition notice is issued by an inspector of the Health and Safety Executive or local authority:

- [ ] A: work can continue, provided that a risk assessment is carried out
- [ ] B: the work that is subject to the notice must cease
- [ ] C: the work can continue if extra safety precautions are taken
- [ ] D: the work in hand can be completed, but no new works started

16.44

When is it advisable to take precautions to prevent the fall of materials into an excavation?

- [ ] A: At all times
- [ ] B: When the excavation is 2 or more metres deep
- [ ] C: When more than five people are working in the excavation
- [ ] D: When there is a risk from an underground cable or other service

16.45

Sole-boards may be omitted if a scaffold is based on a surface:

- [ ] A: of grass
- [ ] B: of sufficient strength
- [ ] C: of a recently backfilled excavation
- [ ] D: of firm mud

Answers: 16.40 = C, 16.41 = C, 16.42 = D, 16.43 = B, 16.44 = A, 16.45 = B

16.46

If a scaffold is not complete, which of the following actions should be taken?

- [ ] A: Make sure the scaffolders complete the scaffold
- [ ] B: Tell all operatives not to use the scaffold
- [ ] C: Use the scaffold with care and display a warning notice
- [ ] D: Prevent access to the scaffold by unauthorised people

16.47

Why is it important that hazards are identified?

- [ ] A: They have the potential to cause harm
- [ ] B: They must all be eliminated before work can start
- [ ] C: They must all be notified to the Health and Safety Executive
- [ ] D: They have to be written on the Health and Safety Law poster

16.48

The Management of Health and Safety at Work Regulations 1999 require risk assessments to be made:

- [ ] A: for all work activities
- [ ] B: when there is a danger of someone getting hurt
- [ ] C: when more than five people are employed
- [ ] D: where an accident has happened previously

16.49

Under the requirements of the Work at Height Regulations, the minimum width of a working platform must be:

- [ ] A: 3 boards wide
- [ ] B: suitable for the job in hand
- [ ] C: 2 boards wide
- [ ] D: any width

16.50

Following a scaffold inspection under the Work at Height Regulations, how soon must a report be given to the person on whose behalf the inspection was made?

- [ ] A: Within 2 hours
- [ ] B: Within 6 hours
- [ ] C: Within 12 hours
- [ ] D: Within 24 hours

Answers: 16.46 = D, 16.47 = A, 16.48 = A, 16.49 = B, 16.50 = D

16.51

If precautions are taken to prevent people and items falling, scaffold guard-rails may be temporarily removed provided that:

- [ ] A: they are replaced as soon as practicable
- [ ] B: the scaffolding does not become overloaded
- [ ] C: the scaffolding does not begin to sway
- [ ] D: the intermediate guard-rail is left in position

16.52

The Work at Height Regulations require a working platform to be inspected:

- [ ] A: after an accident
- [ ] B: every day
- [ ] C: fortnightly
- [ ] D: before first use and then every seven days afterwards

16.53

An assessment has been carried out under the Control of Substances Hazardous to Health Regulations. To which of the following should the risks and control measures be explained?

- [ ] A: All employees on site
- [ ] B: The operatives using the substance
- [ ] C: The person in charge of the stores
- [ ] D: The accounts department

16.54

The Work at Height Regulations require inspections of scaffolding to be carried out by:

- [ ] A: the site manager
- [ ] B: the scaffolder
- [ ] C: the safety adviser
- [ ] D: a competent person

16.55

Under the requirements of the Construction (Design and Management) Regulations, what has to be displayed on a construction site?

- [ ] A: Notice of application to erect hoardings
- [ ] B: Notice of the Health and Safety Commission's address
- [ ] C: Form F10 (Rev)
- [ ] D: A statement by the client

Answers: 16.51 = A, 16.52 = D, 16.53 = B, 16.54 = D, 16.55 = C

16.56

Guidance Notes accompanying regulations are:

- [ ] A: the health and safety rules as laid down by the employer
- [ ] B: a CITB-approved guide book on health and safety
- [ ] C: a set of health and safety guidelines provided by suppliers
- [ ] D: advice on complying with legislation issued by the Health and Safety Executive

16.57

According to Health and Safety Executive figures, most accidents involving site transport are caused by vehicles:

- [ ] A: speeding
- [ ] B: ignoring one-way systems
- [ ] C: reversing
- [ ] D: being used by unqualified drivers

16.58

For a ladder, what is the maximum vertical height that may be climbed before an intermediate landing place is required?

- [ ] A: 7.5 metres
- [ ] B: 8.0 metres
- [ ] C: 8.5 metres
- [ ] D: 9.0 metres

16.59

On a scaffold, the minimum height of the main guard-rail must be:

- [ ] A: 875 mm
- [ ] B: 910 mm
- [ ] C: 950 mm
- [ ] D: 1000 mm

16.60

On a scaffold, the unprotected gap between any guard-rail, toe-board, barrier or other similar means of protection should not exceed:

- [ ] A: 400 mm
- [ ] B: 470 mm
- [ ] C: 500 mm
- [ ] D: 600 mm

16.61

When setting up a fuel storage tank on site, a spillage bund must have a minimum capacity of:

- [ ] A: the contents of the tank + 10%
- [ ] B: the contents of the tank + 15%
- [ ] C: the contents of the tank + 20%
- [ ] D: the contents of the tank + 25%

Answers: 16.56 = D, 16.57 = C, 16.58 = D, 16.59 = C, 16.60 = B, 16.61 = A

16.62

In the context of a risk assessment, what does the term 'risk' mean?

- [ ] A: Something with the potential to cause injury
- [ ] B: An unsafe act or condition
- [ ] C: The likelihood that harm from a particular hazard will be realised
- [ ] D: Any work activity that can be described as dangerous

16.63

Which of the following precautions should be taken to prevent a dumper from falling into an excavation when tipping material into it?

- [ ] A: Keep 5 metres away from the excavation
- [ ] B: Provide a stop block appropriate to the vehicle's wheel size
- [ ] C: Judge the distance carefully
- [ ] D: Approach the excavation in reverse gear

16.64

The number of people who may be carried in a passenger hoist on site must be:

- [ ] A: displayed in the site canteen
- [ ] B: displayed on a legible notice within the cage of the hoist
- [ ] C: given in the company safety policy
- [ ] D: given to the operator of the hoist

16.65

The minimum height of a toe-board on a working platform is:

- [ ] A: 150 mm
- [ ] B: 160 mm
- [ ] C: 170 mm
- [ ] D: 200 mm

16.66

A Health and Safety Inspector may visit your site:

- [ ] A: if 7 days' notice is given
- [ ] B: if any length of notice is given in writing
- [ ] C: at any time without notice
- [ ] D: only if invited by a senior member of staff

Answers: 16.62 = C, 16.63 = B, 16.64 = B, 16.65 = A, 16.66 = C

16.67

The Health and Safety at Work Act and any regulations made under that Act are:

- [ ] A: advisory to companies and individuals
- [ ] B: legally binding
- [ ] C: good practical advice for the employer to follow
- [ ] D: not compulsory, but should be complied with if convenient

16.68

The significant findings of risk assessments must be recorded when more than a certain number of people are employed. How many?

- [ ] A: 3 or more
- [ ] B: 5 or more
- [ ] C: 6 or more
- [ ] D: 7 or more

16.69

What must the planning supervisor ensure is provided at the tendering stage for all demolition projects?

- [ ] A: A construction phase safety plan
- [ ] B: A detailed method statement
- [ ] C: A comprehensive risk assessment
- [ ] D: A pre-tender health and safety plan

16.70

Which two of the following factors must you consider when providing first-aid facilities on site?

- [ ] A: The cost of first-aid equipment
- [ ] B: The hazards and risks that are likely to occur
- [ ] C: The nature of the work carried out
- [ ] D: The difficulty in finding time to purchase the necessary equipment
- [ ] E: The space in the site office to store the necessary equipment

16.71

One of your staff has agreed to become a first-aider and you have been told to make the appropriate arrangements. What course of action should you take?

- [ ] A: Give your volunteer a first-aid manual to read
- [ ] B: Ask a qualified first-aider from another company to brief your volunteer
- [ ] C: Show the volunteer where the first-aid kit is and let them take it from there
- [ ] D: Ensure your volunteer attends an appropriate first-aid course

Answers: 16.67 = B, 16.68 = B, 16.69 = D, 16.70 = B,C, 16.71 = D

16.72

The minimum level of first-aid cover required at any workplace is an appointed person. Which of the following would you expect the appointed person to carry out?

- [ ] A: Provide **most** of the care normally carried out by a first-aider
- [ ] B: Provide **all** the care normally provided by a first-aider
- [ ] C: Contact the emergency services and direct them to the scene of an accident
- [ ] D: Only apply splints to broken bones

16.73

The Health and Safety Information for Employees Regulations require that employees be informed about specific matters of health, safety and welfare in the workplace. In which two of the following ways may this information be communicated to employees?

- [ ] A: By displaying the approved poster
- [ ] B: Verbally during site induction
- [ ] C: By displaying a locally produced poster
- [ ] D: By issuing an approved leaflet
- [ ] E: During a toolbox talk

16.74

Which of the following items of information will you find on the approved Health and Safety Law poster?

- [ ] A: Details of all emergency escape routes
- [ ] B: The identity of the first-aiders
- [ ] C: The location of all fire extinguishers
- [ ] D: The address of the local Health and Safety Executive office

16.75

The advantage of using safety nets rather than harness and lanyard is that safety nets:

- [ ] A: do not need inspecting
- [ ] B: are cheaper
- [ ] C: provide collective fall protection
- [ ] D: can be rigged by anyone

16.76

What should you do if a Health and Safety Inspector arrives on site and wants to see your Site Diary following an accident?

- [ ] A: Hand the diary over
- [ ] B: Tear any pages out that you feel may be incriminating before handing it over
- [ ] C: Flatly refuse to hand over the diary
- [ ] D: Refuse to hand over the diary until you have consulted with Senior Management

Answers: 16.72 = C, 16.73 = A,D, 16.74 = D, 16.75 = D, 16.76 = A

16.77

What should you do for the safety of private motorists if transport leaving your site is likely to deposit mud on the public road?

- [ ] A: Have someone in the road to slow down the traffic
- [ ] B: Employ an on-site method of washing the wheels of site transport
- [ ] C: Employ a mechanical road sweeper
- [ ] D: Have someone hosing-down the mud in the road

16.78

What should you do if you notice that operatives working above a safety net are dropping off-cuts of material and other debris into the net?

- [ ] A: Nothing, as at least it is all collecting in one place
- [ ] B: Ensure that the net is cleared of debris weekly
- [ ] C: Have the net cleared and ensure it is not allowed to happen again
- [ ] D: Ensure that the net is cleared of debris daily

16.79

From a safety point of view, which of the following should be considered first when deciding on the number and location of access and egress points to a site?

- [ ] A: Parking for the supervisor's car
- [ ] B: Access for the emergency services
- [ ] C: Access for heavy vehicles
- [ ] D: Site security

16.80

Which is not classified as a major injury under Reporting of Injuries, Diseases and Dangerous Occurrences Regulations?

- [ ] A: Fractured finger
- [ ] B: Temporary loss of eyesight
- [ ] C: Fractured arm
- [ ] D: Broken wrist

16.81

What should be included in a safety method statement for working at height? Give three answers:

- [ ] A: The cost of the job and time it will take
- [ ] B: The sequence of operations and the equipment to be used
- [ ] C: How much insurance cover will be required
- [ ] D: How falls are to be prevented
- [ ] E: Who will supervise the job on site

Answers: 16.77 = B, 16.78 = C, 16.79 = B, 16.80 = A, 16.81 = B,D,E

16.82

When putting people to work above public areas, your first consideration should be:

- A: minimise the number of people below at any one time
- B: to prevent complaints from the public
- C: to let the public know what you are doing
- D: to prevent anything falling on to people below

16.83

A competent person must routinely inspect a scaffold:

- A: after it is erected and at intervals not exceeding 7 days
- B: only after it has been erected
- C: after it is erected and then at monthly intervals
- D: after it is erected and then at intervals not exceeding 10 days

16.84

Ideally, a safety net should be rigged:

- A: immediately below where you are working
- B: 2 metres below where you are working
- C: 6 metres below where you are working
- D: at any height below the working position

16.85

Someone wearing a safety harness has a fall. What is the main danger of leaving them suspended for too long?

- A: The anchorage point may fail
- B: They may try to climb back up the structure and fall again
- C: They will suffer severe discomfort and may lose consciousness
- D: It will discourage other people from working at height

16.86

Edge protection must be designed to:

- A: make access to the roof easier
- B: secure tools and materials close to the edge
- C: stop rainwater running off the roof onto workers below
- D: prevent people and materials falling

16.87

When should guard-rails be fitted to a working platform?

- A: If it is possible to fall 2 metres
- B: At any height if a fall could result in an injury
- C: If it is possible to fall 3 metres
- D: Only if materials are being stored on the working platform

Answers: 16.82 = D, 16.83 = A, 16.84 = A, 16.85 = C, 16.86 = D, 16.87 = B

16.88

The Beaufort Scale is important when you have people working at height because it measures:

- [ ] A: air temperature
- [ ] B: the load-bearing capacity of a flat roof
- [ ] C: wind speed
- [ ] D: the load-bearing capacity of a scaffold

16.89

A design feature of some air bags used for fall arrest is a controlled leak rate. If you are using these, the inflation pump must:

- [ ] A: be electrically-powered
- [ ] B: be switched off from time to time to avoid over-inflation
- [ ] C: run all the time while work is carried out at height
- [ ] D: switched off when the air bags are full

16.90

Why is it dangerous to use inflatable air bags that are too big for the area to be protected?

- [ ] A: They will exert a sideways pressure on anything that is containing them
- [ ] B: The pressure in the bags will cause them to burst
- [ ] C: The inflation pump will become overloaded
- [ ] D: They will not fully inflate

16.91

Who should you inform if someone reports to you that they have work-related hand-arm vibration syndrome?

- [ ] A: The Health and Safety Executive
- [ ] B: The local Health Authority
- [ ] C: A coroner
- [ ] D: The nearest hospital

16.92

How should cylinders containing Liquid Petroleum Gas be stored on site?

- [ ] A: In a locked cellar with clear warning signs
- [ ] B: In a locked external compound at least 3 metres from any oxygen cylinders
- [ ] C: As close to the point of use as possible
- [ ] D: Covered by a tarpaulin to shield the compressed cylinder from sunlight

16.93

How should access be controlled, if you have people working in a riser shaft?

- [ ] A: By a site security operative
- [ ] B: By those who are working in it
- [ ] C: By the main contractor
- [ ] D: By a Permit to work system

Answers: 16.88 = C, 16.89 = C, 16.90 = A, 16.91 = A, 16.92 = B, 16.93 = D

16.94

Before allowing a lifting operation to be carried out, you must ensure that the sequence of operations to enable a lift to be carried out safely is confirmed in:

- [ ] A: verbal instructions
- [ ] B: a method statement
- [ ] C: a radio telephone message
- [ ] D: a notice in the canteen

16.95

Where should liquefied petroleum gas (Liquid Petroleum Gas) cylinders be positioned when supplying an appliance in a site cabin?

- [ ] A: Inside the site cabin in a locked cupboard
- [ ] B: Under the cabin
- [ ] C: Inside the cabin next to the appliance
- [ ] D: Outside the cabin

16.96

Welding is about to start on your site. What should be provided to prevent passers-by from getting arc-eye:

- [ ] A: Warning signs
- [ ] B: Screens
- [ ] C: Personal Protective Equipment
- [ ] D: Nothing

16.97

What must a subcontractor provide you with in relation to one of his employees who is 17 years old?

- [ ] A: The employee's birth certificate
- [ ] B: Health and Safety Executive permission for the 17 year old be on site
- [ ] C: Parental permission for the 17 year old be on site
- [ ] D: A risk assessment addressing the issue of young persons

16.98

Why may young people be more at risk of having accidents?

- [ ] A: Legislation does not apply to anyone under 18 years of age
- [ ] B: They are usually left to work alone to gain experience
- [ ] C: They are inexperienced and may not recognise danger
- [ ] D: There is no legal duty to provide them with Personal Protective Equipment

Answers: 16.94 = B, 16.95= D, 16.96 = B, 16.97 = D, 16.98 = C

16.99

What is your least reliable source of information when assessing the level of vibration from a powered percussive hand tool?

- [ ] A: In-use vibration measurement of the tool
- [ ] B: Vibration figures taken from the tool manufacturer's hand book
- [ ] C: Your own judgment based upon observation
- [ ] D: Vibration data from the Health and Safety Executive's master list

Answer: 16.99 = C

17 Demolition

17.1

Every demolition contractor undertaking demolition operations must first appoint:

- [ ] A: a competent person to supervise the work
- [ ] B: a sub-contractor to strip out the buildings
- [ ] C: a safety officer to check on health and safety compliance
- [ ] D: a quantity surveyor to price the extras

17.2

If there are any doubts as to a building's stability, a demolition contractor should consult:

- [ ] A: another demolition contractor
- [ ] B: a structural engineer
- [ ] C: an Health and Safety Executive Factory Inspector
- [ ] D: the company safety adviser

17.3

Which two of the following documents refer to the specific hazards associated with demolition work in confined spaces?

- [ ] A: A safety policy
- [ ] B: A permit-to-work
- [ ] C: A risk assessment
- [ ] D: A scaffolding permit
- [ ] E: The Health and Safety Executive Health and Safety Law poster

17.4

What should you do if you discover underground services not previously identified?

- [ ] A: Fill in the hole and say nothing to anyone
- [ ] B: Stop work until the situation has been resolved
- [ ] C: Cut the pipe or cable to see if it's live
- [ ] D: Get the machine driver to dig it out

17.5

Continual use of hand-held breakers or drills is most likely to cause:

- [ ] A: dermatitis
- [ ] B: Weil's disease
- [ ] C: vibration white finger
- [ ] D: skin cancer

17.6

With regard to the safe method of working, what is the most important subject of induction training for demolition operatives?

- [ ] A: Working hours on the site
- [ ] B: Explanation of the method statement
- [ ] C: Location of welfare facilities
- [ ] D: COSHH assessments

Answers: 17.1 = A, 17.2 = B, 17.3 = B,C, 17.4 = B, 17.5 = C, 17.6 = B

17.7

What is the most common source of high levels of lead in the blood of operatives doing demolition work?

- [ ] A: Stripping lead sheeting
- [ ] B: Cutting lead-covered cable
- [ ] C: Cold cutting fuel tanks
- [ ] D: Hot cutting coated steel

17.8

When asbestos material is suspected in buildings to be demolished, what is the first priority?

- [ ] A: A competent person carries out an asbestos survey
- [ ] B: Notify the Health and Safety Executive of the possible presence of asbestos
- [ ] C: Remove and dispose of the asbestos
- [ ] D: Employ a licensed asbestos remover

17.9

On site, what is the minimum distance that oxygen should be stored away from propane, butane or other gases?

- [ ] A: 1 metre
- [ ] B: 2 metres
- [ ] C: 3 metres
- [ ] D: 4 metres

17.10

Where should Liquid Petroleum Gas cylinders be located when being used for heating or cooking in site cabins?

- [ ] A: Under the kitchen area work surface
- [ ] B: Inside but near the door for ventilation
- [ ] C: In a nearby storage container
- [ ] D: Outside the cabin

17.11

What type of fire extinguisher should not be provided where petrol is being stored?

- [ ] A: Foam
- [ ] B: Water
- [ ] C: Dry powder
- [ ] D: Carbon dioxide

17.12

What action should you take if you discover unlabelled drums or containers on site?

- [ ] A: Put them in the nearest waste skip
- [ ] B: Ignore them. They will get flattened during the demolition
- [ ] C: Stop work until they have been safely dealt with
- [ ] D: Open them and smell the contents

Answers: 17.7 = D, 17.8 = A, 17.9 = C, 17.10 = D, 17.11 = B, 17.12 = C

17.13

What is the safest way of identifying non-load bearing walls in a building?

- [ ] A: Look at structural plans
- [ ] B: Break out a test hole
- [ ] C: Investigate above ceiling level
- [ ] D: Ask the previous occupants

17.14

Which is the safest method of demolishing brick or internal walls by hand?

- [ ] A: Undercut the wall at ground level
- [ ] B: Work across in even courses from the ceiling down
- [ ] C: Work from the doorway at full height
- [ ] D: Cut down at corners and collapse in sections

17.15

Who should be consulted before demolition is carried out near to overhead cables?

- [ ] A: The Health and Safety Executive
- [ ] B: The Fire Service
- [ ] C: The electricity supply company
- [ ] D: The land owner

17.16

When demolishing a building in controlled sections, what is the most important consideration for the remaining structure?

- [ ] A: The soft strip is completed
- [ ] B: All non-ferrous metals are removed
- [ ] C: That it remains stable
- [ ] D: Trespassers cannot get in at night

17.17

Which of the following items of Personal Protective Equipment provides the lowest level of protection when working in dusty conditions?

- [ ] A: Half-mask dust respirator
- [ ] B: Positive pressure powered respirator
- [ ] C: Compressed airline breathing apparatus
- [ ] D: Self-contained breathing apparatus

Answers: 17.13 = A, 17.14 = B, 17.15 = C, 17.16 = C, 17.17 = A

17.18

Which two of the following would be suitable to use when cutting coated steelwork?

- [ ] A: A disposable dust-mask
- [ ] B: Positive pressure powered respirator
- [ ] C: High-efficiency dust respirator
- [ ] D: Ventilated helmet respirator
- [ ] E: Respiratory protection is not required

17.19

Where would you find the intended method of controlling identified hazards on a demolition project?

- [ ] A: The structural plans
- [ ] B: The risk assessments
- [ ] C: The site welfare plans
- [ ] D: The Health and Safety Law poster

17.20

What method of storing cans or drums must be used to prevent any leakage spreading?

- [ ] A: On wooden pallets off the ground
- [ ] B: On their sides and chocked to prevent movement
- [ ] C: Upside down to prevent water penetrating the screw top
- [ ] D: In a bund in case of leakage

17.21

Before entering large open-topped tanks, what is the most important thing you should obtain?

- [ ] A: A ladder for easy access
- [ ] B: A valid permit to work
- [ ] C: An operative to keep watch over you
- [ ] D: A gas meter to detect any gas

17.22

After exposure to lead, what precautions should you take before eating or drinking?

- [ ] A: Wash your hands and face
- [ ] B: Do not smoke
- [ ] C: Change out of dirty clothes
- [ ] D: Brush your teeth

17.23

When hinge-cutting a steel building or structure for a 'controlled collapse', which should be the last cuts?

- [ ] A: Front leading row top cuts
- [ ] B: Front leading row bottom cuts
- [ ] C: Back-row top cuts
- [ ] D: Back-row bottom cuts

Answers: 17.18 = B,D, 17.19 = B, 17.20 = D, 17.21 = B, 17.22 = A, 17.23 = D

17.24

What safety devices should be fitted between the pipes and the gauges of oxy-propane cutting equipment?

- [ ] A: Non-return valves
- [ ] B: On-off taps
- [ ] C: Flame retardant tape
- [ ] D: Flashback arresters

17.25

What type of fire extinguisher should you not use in confined spaces?

- [ ] A: Water
- [ ] B: Carbon dioxide
- [ ] C: Dry powder
- [ ] D: Foam

17.26

Before carrying out the demolition cutting of fuel tanks what should be obtained?

- [ ] A: A gas free certificate
- [ ] B: An isolation certificate
- [ ] C: A risk assessment
- [ ] D: A COSHH assessment

17.27

How long is a gas free certificate issued for?

- [ ] A: One week
- [ ] B: One month
- [ ] C: One day
- [ ] D: One hour

17.28

Which asbestos material is classified as a 'non-notifiable' for removal works:

- [ ] A: Asbestos cement
- [ ] B: Asbestos insulation / coatings
- [ ] C: Asbestos insulation board
- [ ] D: Asbestos pipe lagging

17.29

What do the letters SWL stand for?

- [ ] A: Satisfactory working limit
- [ ] B: Safe working level
- [ ] C: Satisfactory weight limit
- [ ] D: Safe working load

Answers: 17.24 = D, 17.25 = B, 17.26 = A, 17.27 = C, 17.28 = A, 17.29 = D

17.30

Which of the following is true as regards the safe working load of a piece of equipment?

- [ ] A: It must never be exceeded
- [ ] B: It is a guide figure that may be exceeded slightly
- [ ] C: It may be exceeded by 10% only
- [ ] D: It gives half the maximum weight to be lifted

17.31

What should be clearly marked on all lifting gear?

- [ ] A: Date of manufacture
- [ ] B: Name of maker
- [ ] C: Date next due for test
- [ ] D: Safe working load

17.32

Which of the following pieces of equipment can a 17-year-old trainee demolition operative use unsupervised?

- [ ] A: Excavator 360 degrees
- [ ] B: Dump truck
- [ ] C: Wheelbarrow
- [ ] D: Rough terrain forklift

17.33

Lifting accessories must be thoroughly examined every:

- [ ] A: 3 months
- [ ] B: 6 months
- [ ] C: 14 months
- [ ] D: 18 months

17.34

Plant and equipment needs to be inspected and the details recorded by operators:

- [ ] A: daily at the beginning of each shift
- [ ] B: weekly
- [ ] C: monthly
- [ ] D: every 3 months

17.35

What is the importance of having ROPS fitted to some mobile plant?

- [ ] A: It ensures that the pressure of the tyres is correct
- [ ] B: It protects the operator if the machine rolls over
- [ ] C: It prevents over-pressurisation of the hydraulic system
- [ ] D: It prevents un-authorised passengers being carried

Answers: 17.30 = A, 17.31 = D, 17.32 = C, 17.33 = B, 17.34 = B, 17.35 = B

17.36

The plant you are driving has defective brakes. What action should you take?

- A: Reduce your speed accordingly
- B: Report it and carry on working
- C: Report it and isolate the machine
- D: Use the handbrake until the machine fixed

17.37

The correct way to climb off a machine is to:

- A: jump down from the seated position
- B: climb down, facing forward
- C: climb down, facing the machine
- D: use a ladder

17.38

It is acceptable to carry passengers on a machine provided:

- A: the employer gives permission
- B: they are carried in the skip
- C: a purpose-made seat is provided
- D: the maximum speed is no greater than 10 mph

17.39

Where do you find information about daily checks required for mobile plant?

- A: On the stickers attached to the machine
- B: In the manufacturer's handbook
- C: In the supplier's information
- D: Any or all of the other answers

17.40

What action should be taken if a wire rope sling is defective?

- A: Do not use it and make sure that no one else can
- B: Only use it for up to half its safe working load
- C: Put it to one side to wait for repair
- D: Only use it for small lifts under 1 tonne

17.41

What should you do while reversing mobile plant if you lose sight of the signaller or slinger who is directing you?

- A: Carry on reversing slowly
- B: Stop the vehicle
- C: Adjust your wing mirror
- D: Sound the horn and move forward

Answers: 17.36 = C, 17.37 = C, 17.38 = C, 17.39 = D, 17.40 = A, 17.41 = B

17.42

How often should lifting equipment that is not used to lift people be thoroughly examined?

- [ ] A: Once every 6 months
- [ ] B: At least twice a year
- [ ] C: A minimum of once a year
- [ ] D: Every 10 years

17.43

When leaving mobile plant unattended, you should:

- [ ] A: leave the engine running if safe to do so
- [ ] B: park it in a safe place, remove the keys and lock it
- [ ] C: put the parking brake on and tell no one to use it
- [ ] D: put a sign saying 'No unauthorised access' on it

17.44

Which statement is true with regard to using machines?

- [ ] A: Guards can be removed to make work easier
- [ ] B: It's OK to wear rings and other jewellery so long as you take care
- [ ] C: Carefully remove waste material while the machine is in motion
- [ ] D: Never use a machine unless you have been trained and given permission to use it

17.45

Which of the following is not a part of a plant operators daily pre-use check?

- [ ] A: Emergency systems
- [ ] B: Engine oil level
- [ ] C: Hydraulic fluid level
- [ ] D: Brake pad wear

17.46

An operator of a scissor lift must:

- [ ] A: be trained and authorised in the use of the equipment
- [ ] B: only use the ground level controls
- [ ] C: be in charge of the work team
- [ ] D: ensure that only one person is on the platform at a time

17.47

The operation of a scissor lift would be unsafe if:

- [ ] A: the controls on the platform are used
- [ ] B: the ground is soft and sloping
- [ ] C: weather protection is not fitted
- [ ] D: the machine is short of fuel

Answers: 17.42 = C, 17.43 = B, 17.44 = D, 17.45 = D, 17.46 = A, 17.47 = B

17.48

On demolition sites, what must the drivers of plant have, for their own and others' safety?

- [ ] A: Adequate visibility from the driving position
- [ ] B: A temperature controlled cab
- [ ] C: Wet weather gear when it's raining
- [ ] D: A supervisor in the cab with them

17.49

When must head and tail lights be used on mobile plant?

- [ ] A: Only if using the same traffic route as private cars
- [ ] B: In all conditions of poor visibility
- [ ] C: When operated by a trainee
- [ ] D: Only if crossing pedestrian routes

17.50

With regard to mobile plant, what safety feature is provided by FOPS?

- [ ] A: The speed is limited when tracking over hard surfaces
- [ ] B: The machine stops automatically if the operator lets go of the controls
- [ ] C: The operator is protected from falling materials
- [ ] D: The reach is limited when working near to live overhead cables

Answers: 17.48 = A, 17.49 = B, 17.50 = C

18.1

You are working in an occupied building and have taken up 6 lengths of 3-metre floor boarding when you are called away to an urgent job. What should you do?

- [ ] A: Leave the job as it is
- [ ] B: Cordon off the work area before leaving the job
- [ ] C: Permanently re-fix the floorboards and floor coverings
- [ ] D: Tell other workers to be careful while you are away

18.2

You are preparing to use an electric powered threading machine. Which of the following statements should apply?

- [ ] A: The power supply should be 24 volts and the machine fitted with a guard
- [ ] B: The power supply should be 415 volts and the machine fitted with a guard
- [ ] C: Ensure your clothing cannot get caught on rotating parts of the machine
- [ ] D: Ear defenders should be available and should be in good condition

18.3

In terms of accident prevention, which two of the following are the most important precautions when working in the roof space of an occupied house?

- [ ] A: Safe foot access over the ceiling joists
- [ ] B: Safe access into the roof space using a stepladder or ladder
- [ ] C: Check that the roof space is free from hornets' nests
- [ ] D: The roof space must be fitted with a light socket and switch
- [ ] E: Carry out a test to see if the plasterboard ceiling will support your weight

18.4

You have removed a WC pan in a public toilet and notice a hypodermic syringe lodged in the soil pipe connector. What should you do?

- [ ] A: Ensure the syringe is empty, remove the syringe and place it with the rubbish
- [ ] B: Wear gloves, break the syringe into small pieces and flush it down the drain
- [ ] C: Notify the supervisor, cordon the area off and call the emergency services
- [ ] D: Wearing gloves, use grips to remove the syringe to a safe place and tell your supervisor

Answers: 18.1 = B, 18.2 = C, 18.3 = A, B, 18.4 = D

18.5

You have been handling sheet lead. How is some lead most likely to get in your bloodstream?

- [ ] A: By not using the correct respirator
- [ ] B: By not washing your hands before eating
- [ ] C: By not changing out of your work clothes
- [ ] D: By not wearing safety goggles

18.6

The legionella bacteria that cause Legionnaire's disease are most likely to be found in which of the following?

- [ ] A: A boiler operating at a temperature of 80° centigrade
- [ ] B: A shower hose outlet
- [ ] C: A cold water storage cistern containing water at 10° centigrade
- [ ] D: A WC toilet pan

18.7

How are legionella bacteria passed on to humans?

- [ ] A: Through fine water droplets such as sprays or mists
- [ ] B: By drinking dirty water
- [ ] C: Through contact with the skin
- [ ] D: From other people when they sneeze

18.8

When bossing a sheet lead corner using lead working tools, you:

- [ ] A: are allowed to smoke during the bossing process
- [ ] B: should not smoke during the bossing process
- [ ] C: should only smoke when your hands are protected with barrier cream
- [ ] D: are allowed to smoke if you are wearing gloves

18.9

When replacing an electrical immersion heater in a hot water storage cylinder, what should you do to make sure that the electrical supply is dead before starting plumbing work?

- [ ] A: Switch off and disconnect the supply to the immersion heater
- [ ] B: Switch off and cut through the electric cable with insulated pliers
- [ ] C: Switch off and test the circuit
- [ ] D: Switch off, isolate the supply at the mains board and test the circuit

Answers: 18.5 = B, 18.6 = B, 18.7 = A, 18.8 = B, 18.9 = D

18.10

The reason for carrying out temporary continuity bonding before removing and replacing sections of metallic pipework is to:

- [ ] A: provide a continuous earth for the pipework installation
- [ ] B: prevent any chance of blowing a fuse
- [ ] C: maintain the live supply to the electrical circuit
- [ ] D: prevent any chance of corrosion to the pipework

18.11

You are required to re-fix a section of external rainwater pipe using a power drill in wet weather conditions. Which type of drill is most suitable?

- [ ] A: Battery-powered drill
- [ ] B: Drill with 110 volt power supply
- [ ] C: Drill with 240 volt power supply
- [ ] D: Any mains voltage drill with a power breaker

18.12

What piece of equipment would you use to find out whether a section of solid wall that you are about to drill into contains electric cables?

- [ ] A: A neon screwdriver
- [ ] B: A cable tracer
- [ ] C: A multimeter
- [ ] D: A hammer and chisel

18.13

When maintaining or installing a central heating pump in a domestic property, the correct electricity supply to the pump should be:

- [ ] A: 24 volts
- [ ] B: 110 volts
- [ ] C: 240 volts
- [ ] D: 415 volts

18.14

Why is it important that operatives know the difference between propane and butane equipment?

- [ ] A: Propane equipment operates at higher pressure
- [ ] B: Propane equipment operates at lower pressure
- [ ] C: Propane equipment is cheaper
- [ ] D: Propane equipment can be used with smaller, easy-to-handle cylinders

18.15

Which of the following statements is true?

- [ ] A: Both propane and butane are heavier than air
- [ ] B: Butane is heavier than air while propane is lighter than air
- [ ] C: Propane is heavier than air while butane is lighter than air
- [ ] D: Both propane and butane are lighter than air

Answers: 18.10 = A, 18.11 = A, 18.12 = B, 18.13 = C, 18.14 = A, 18.15 = A

18.16

Apart from the cylinders used in gas-powered forklift trucks, liquefied petroleum gas cylinders should never be placed on their sides during use because:

- [ ] A: it would give a faulty reading on the contents gauge, resulting in flashback
- [ ] B: air could be drawn into the cylinder, creating a dangerous mixture of gases
- [ ] C: the liquid gas would be at too low a level to allow the torch to burn correctly
- [ ] D: the liquid gas could be drawn from the cylinder, creating a safety hazard

18.17

What is the preferred method of checking for leaks when assembling liquefied petroleum gas equipment before use?

- [ ] A: Test with a lighted match
- [ ] B: Sniff the connections to detect the smell of gas
- [ ] C: Listen to hear for escaping gas
- [ ] D: Apply leak detection fluid to the connections

18.18

When transporting liquefied petroleum gas cylinders (above 5 kg) in an enclosed van, under the Packaged Goods Regulations the driver must:

- [ ] A: be trained and competent in hazards relating to liquefied petroleum gas
- [ ] B: have a heavy goods vehicle licence
- [ ] C: under no circumstances carry liquefied petroleum gas cylinders
- [ ] D: have a full driving licence with a Packaged Goods Regulations endorsement

18.19

What is the colour of propane gas cylinders?

- [ ] A: Black
- [ ] B: Maroon
- [ ] C: Red / Orange
- [ ] D: Blue

Answers: 18.16 = D, 18.17 = D, 18.18 = A, 18.19 = C

18.20

Which of the following statements is true? It is safe to transport employees to the workplace in the rear of a van if:

- [ ] A: the driver has a heavy goods vehicle licence
- [ ] B: the van is fitted with temporary seating
- [ ] C: the van is fitted with proper seating and seat belts
- [ ] D: the driver has a public service vehicle licence

18.21

Which is the safest method of taking long lengths of copper pipe by van?

- [ ] A: Tying the pipes to the roof with copper wire
- [ ] B: Someone holding the pipes on the roof rack as you drive along
- [ ] C: Putting the pipes inside the van with the ends out of the passenger window
- [ ] D: Using a pipe rack fixed to the roof of the van

18.22

When should you wear a safety belt when driving a dumper truck on site?

- [ ] A: When travelling on rough ground
- [ ] B: When the dumper is loaded
- [ ] C: When carrying passengers
- [ ] D: Whenever one is provided

18.23

When using a blowtorch to joint copper tube and fittings in a domestic property, a fire extinguisher should be:

- [ ] A: available in the immediate work area
- [ ] B: held over the joint while you are using the blowtorch
- [ ] C: used to cool the fitting
- [ ] D: available only if a property is occupied

Answers: 18.20 = C, 18.21 = D, 18.22 = D, 18.23 = A

18.24

When using a blowtorch, you should:

- A: stop using the blowtorch immediately before leaving the job
- B: stop using the blowtorch at least 1 hour before leaving the job
- C: stop using the blowtorch at least 2 hours before leaving the job
- D: stop using the blowtorch at least 4 hours before leaving the job

18.25

When using a blowtorch near to hair-felt lagging you should:

- A: just remove enough lagging carry out the work
- B: remove the lagging at least 1 metre either side of the work
- C: remove the lagging at least 3 metres either side of the work
- D: wet the lagging but leave it in place

18.26

When using a blowtorch near to timber, you should:

- A: carry out the work taking care not to catch the timber
- B: use a non-combustible mat and have a fire extinguisher ready
- C: wet the timber first and have a bucket of water handy
- D: point the flame away from the timber and have a bucket of sand ready to put out the fire

18.27

You are fixing sheet lead flashing to a chimney on the roof of a busy town centre shop. What is the most important thing you must do?

- A: Pull the ladder onto the roof to prevent the public from climbing up
- B: Make provision for protecting the public from objects that could fall
- C: Wear a face mask to protect you from breathing the chimney fumes
- D: Wear safety boots to prevent you from slipping off the roof

Answers: 18.24 = B, 18.25 = B, 18.26 = B, 18.27 = B

18.28

Stepladders must only be used:

- [ ] A: inside buildings
- [ ] B: if they are in good condition and suitable
- [ ] C: if they are made of aluminium
- [ ] D: if they are less than 1.75 metres high

18.29

You are removing guttering from a large, single-storey, metal-framed and cladded building. The job is likely to take all day. What is the most appropriate type of access equipment you could use?

- [ ] A: A ladder
- [ ] B: A mobile tower scaffold
- [ ] C: A putlog scaffold
- [ ] D: A trestle scaffold

18.30

You arrive at a job which involves using ladder access to the roof. You notice the ladder has been painted. You should:

- [ ] A: only use the ladder if it is made of metal
- [ ] B: only use the ladder if it is made of wood
- [ ] C: only use the ladder if wearing rubber-soled boots to prevent slipping
- [ ] D: not use the ladder, and report the matter to your supervisor

18.31

If lifting a roll of Code 5 sheet lead, what is the first thing you should do?

- [ ] A: Weigh the roll of lead
- [ ] B: Have a trial lift to see how heavy it feels
- [ ] C: Assess the whole task
- [ ] D: Ask your workmate to give you a hand

18.32

When dismantling lengths of cast iron soil pipe at height, you should:

- [ ] A: work in pairs, break the length at the collar and remove the pipe section
- [ ] B: tie a rope to the pipe and pull the section down. Shout 'below' to warn others
- [ ] C: crack the pipe at the joint and push it away, making sure the area is clear
- [ ] D: smash the pipe into sections using a hammer and remove it piece by piece

Answers: 18.28 = B, 18.29 = B, 18.30 = D, 18.31 = C, 18.32 = A

18.33

You need to move a cast iron bath which is too heavy to lift. You should:

- [ ] A: inform your supervisor and ask for assistance
- [ ] B: get some lifting tackle
- [ ] C: give it another try
- [ ] D: try and find someone to give you a hand

18.34

When installing a flue liner in an existing chimney, you should:

- [ ] A: insert the liner from ground level, pushing it up the chimney
- [ ] B: work in pairs and insert the liner from roof level, working off the roof
- [ ] C: work in pairs and insert the liner from roof level, working from roof ladders or a chimney scaffold
- [ ] D: break through the chimney in the loft area and insert the liner from there

18.35

You are asked to move a cast iron boiler some distance. What should you do?

- [ ] A: Get a workmate to carry it
- [ ] B: Drag it
- [ ] C: Roll it end-over-end
- [ ] D: Use a suitable trolley or other manual handling aid

18.36

If working on a plumbing job where noise levels are rather high, who would you expect to carry out noise assessment?

- [ ] A: A fully qualified plumber
- [ ] B: Your supervisor
- [ ] C: The site engineer
- [ ] D: A competent person

18.37

If breaking up a cast iron bath, which of the following is the proper way to protect your hearing?

- [ ] A: A portable stereo and head set
- [ ] B: Cotton wool pads
- [ ] C: Ear defenders
- [ ] D: Rolled-up tissue paper

Answers: 18.33 = A, 18.34 = C, 18.35 = D, 18.36 = D, 18.37 = C

18.38

You notice that the handle of your wooden 'rat tail' file is starting to split. What should you do?

- [ ] A: Wrap it tightly with plastic adhesive tape
- [ ] B: Use it until the handle falls off completely and then replace the handle
- [ ] C: Remove the damaged handle and replace it
- [ ] D: Throw the damaged tool away

18.39

What is the most likely risk of injury when cutting large diameter pipe?

- [ ] A: Your fingers may become trapped between the cutting wheel and the pipe
- [ ] B: The inside edge of the cut pipe becomes extremely sharp to touch
- [ ] C: Continued use can cause muscle damage
- [ ] D: Pieces of sharp metal could fly off

18.40

The use of oxyacetylene equipment is not recommended for which of the following jointing methods?

- [ ] A: Jointing copper pipe using hard soldering
- [ ] B: Jointing copper tube using capillary soldered fittings
- [ ] C: Jointing mild steel tube
- [ ] D: Jointing sheet lead

18.41

Which of the following makes it essential to take great care when handling oxygen cylinders?

- [ ] A: They contain highly flammable compressed gas
- [ ] B: They contain highly flammable liquid gas
- [ ] C: They are filled to extremely high pressures
- [ ] D: They contain poisonous gas

18.42

What is the colour of an acetylene cylinder?

- [ ] A: Orange
- [ ] B: Black
- [ ] C: Green
- [ ] D: Maroon

Answers: 18.38 = C, 18.39 = B, 18.40 = B, 18.41 = C, 18.42 = D

18.43

What item of personal protective equipment, from the following list, should you use when oxyacetylene welding?

- [ ] A: Ear defenders
- [ ] B: Clear goggles
- [ ] C: Green-tinted goggles
- [ ] D: Dust mask

18.44

When using oxyacetylene welding equipment, the bottles should be

- [ ] A: laid on their side
- [ ] B: stood upright
- [ ] C: stood upside down
- [ ] D: angled at 45°

18.45

Which of the following is the safest place to store oxyacetylene gas welding bottles when they are not in use?

- [ ] A: Outside in a special storage compound
- [ ] B: In company vehicles
- [ ] C: Inside the building in a locked cupboard
- [ ] D: In the immediate work area, ready for use the next day

18.46

You are drilling a hole through a metal partition to receive a 15 mm pipe from a radiator. You need to wear eye protection when

- [ ] A: drilling overhead only
- [ ] B: the drill bit exceeds 20 mm
- [ ] C: always, whatever the circumstances
- [ ] D: drilling through concrete only

18.47

You are asked to join plastic soil pipes in the roof space of a building using a strong-smelling hazardous solvent but you have not been provided with any respiratory protective equipment. What should you do?

- [ ] A: Just get on and do the job
- [ ] B: Use a dust mask
- [ ] C: Sniff the solvent to see if it has any ill-effects on you
- [ ] D: Stop what you are doing and get out

18.48

In plumbing work, what part of the body could suffer long-term damage when hand bending copper pipe using an internal spring?

- [ ] A: Elbows
- [ ] B: Hands
- [ ] C: Back
- [ ] D: Knees

Answers: 18.43 = C, 18.44 = B, 18.45 = A, 18.46 = C, 18.47 = D, 18.48 = D

18.49

When repairing a burst water main using pipe-freezing equipment to isolate the damaged section of pipe, you should:

- [ ] A: always work in pairs when using pipe-freezing equipment
- [ ] B: never allow the freezing gas to come into direct contact with surface water
- [ ] C: never use pipe-freezing equipment on plastic pipe
- [ ] D: wear gloves to avoid direct contact with the skin

18.50

You are drilling a 100 mm diameter hole for a flue pipe through a brick wall with a large hammer drill. Which combination of Personal Protective Equipment (PPE) should you be supplied with?

- [ ] A: Gloves, breathing apparatus and boots
- [ ] B: Ear defenders, face mask and boots
- [ ] C: Ear defenders, breathing apparatus and barrier cream
- [ ] D: Barrier cream, boots and face mask

18.51

During a refurbishment job, you may need to work under a ground-level suspended timber floor. What is the first question you should ask?

- [ ] A: Can the work be performed from outside?
- [ ] B: Will temporary lighting be used?
- [ ] C: What is contained under the floor?
- [ ] D: How many ways in or out are there?

18.52

You are required to replace below-ground drainage pipework in an excavation, which is approximately 2.5 metres deep. The trench sides show signs of collapse. What do you do?

- [ ] A: Get on with the work as quickly as possible
- [ ] B: Refuse to do the work until the trench sides have been properly shored
- [ ] C: Get a mate to help you so that they can pass the materials down to you
- [ ] D: Ensure that you do the work with a rope around you so that you can be pulled out

Answers: 18.49 = D, 18.50 = B, 18.51 = A, 18.52 = B

18.53

When working with fibreglass roof insulation, which of the following items of personal protective equipment (PPE) should you wear?

- [ ] A: Gloves, face mask and eye protection
- [ ] B: Boots, eye protection and ear defenders
- [ ] C: Ear defenders, face mask and boots
- [ ] D: Barrier cream, eye protection and face mask

18.54

While removing old sheet lead from a roof, you find a white powdery substance on the wooden surface beneath the sheet lead. What should you do?

- [ ] A: Carefully sweep off the powder and wash down the surface with soapy water
- [ ] B: Before undertaking any further work, report the matter to your supervisor
- [ ] C: Remove the substance wearing a face mask, and place it in a bag for disposal
- [ ] D: Call in a health and safety specialist to clean up the substance

18.55

When working on a central heating system, you come across some pipework insulated with a hard white powdery material that could be asbestos. What should you do?

- [ ] A: While wearing a face mask, remove the material and dispose of it safely
- [ ] B: Remove the material, putting it back on the pipework after finishing the job
- [ ] C: Stop work immediately and tell your supervisor about the material
- [ ] D: Damp the material down with water and remove it before carrying out the work

18.56

While climbing a ladder, a colleague slips and falls about 2 metres. They are lying on the ground saying that their back hurts. What is the first thing that you should do?

- [ ] A: Go and put it in the accident book
- [ ] B: Help them get up
- [ ] C: Tell them to lie still and send someone else to get a first-aider
- [ ] D: Make sure there is nothing wrong with the ladder

Answers: 18.53 = A, 18.54 = B, 18.55 = C, 18.56 = C

19.1

When carrying out kerbing works, which method should be used for getting kerbs off the vehicle?

- [ ] A: Lift them off manually using the correct technique
- [ ] B: Push them off the back
- [ ] C: Use mechanical means, such as a JCB fitted with a grab
- [ ] D: Ask your workmate to give you a hand

19.2

What are two effects of under-inflated tyres on the operation of a machine?

- [ ] A: It decreases the operating speed of the engine
- [ ] B: It leads to instability of the machine
- [ ] C: It causes increased tyre wear
- [ ] D: It decreases tyre wear
- [ ] E: It increases the operating speed of the engine

19.3

If you are driving any plant which may overturn, when must a seat belt be worn?

- [ ] A: When travelling over rough ground
- [ ] B: When the vehicle is loaded
- [ ] C: When you are carrying passengers
- [ ] D: At all times

19.4

What precautions should be taken when parking on a gradient with the front of the vehicle pointing down the hill?

- [ ] A: Engine off, turn wheels onto full lock
- [ ] B: Hand-brake on, engine off and remove the key
- [ ] C: Leave the engine running and park crosswise
- [ ] D: Apply the hand-brake and put chocks under the wheels

19.5

If the dead man's handle on a machine does not operate, what should you do?

- [ ] A: Keep quiet in case you get the blame
- [ ] B: Report it at the end of the shift
- [ ] C: Try and fix it or repair it yourself
- [ ] D: Stop and report it immediately to your supervisor

19.6

You may carry passengers on vehicles:

- [ ] A: If your supervisor gives you permission
- [ ] B: Only if a suitable secure seat is provided for each of them
- [ ] C: Only when off the public highway
- [ ] D: If you have a full driving licence

Answers: 19.1 = C, 19.2 = B,C, 19.3 = D, 19.4 = B, 19.5 = D, 19.6 = B

19.7

When tipping into an excavation is necessary, what is the preferred method of preventing the vehicle getting too close to the edge?

- [ ] A: Signage
- [ ] B: Driver's experience
- [ ] C: Signaller / Banksman
- [ ] D: Stop-blocks

19.8

Which of these is the safest method of operating a machine which has an extending jib or boom under electrical power lines?

- [ ] A: Erect signs to warn drivers while they are operating the machines
- [ ] B: Adapt the machines to limit the extension of jib or boom
- [ ] C: Place a warning notice on the machine
- [ ] D: Let down the tyres on the machine to increase the clearance

19.9

Which of the following is true as regards the safe working load of lifting equipment such as a cherry picker, lorry loader or excavator?

- [ ] A: It must never be exceeded
- [ ] B: It is a guide figure that may be exceeded slightly
- [ ] C: It may be exceeded by 10% only
- [ ] D: It gives half the maximum weight to be lifted

19.10

Which of the following checks should the operator of a mobile elevating work platform, for example a cherry picker, carry out before using it?

- [ ] A: Check that a seat belt is provided for the operator
- [ ] B: Check that a roll-over cage is fitted
- [ ] C: Drain the hydraulic system
- [ ] D: Check that emergency systems operate correctly

Answers: 19.7 = D, 19.8 = B, 19.9 = A, 19.10 = D

19.11

In which of the following circumstances would it not be safe to use a cherry picker for working at height?

- [ ] A: When a roll-over cage is not fitted
- [ ] B: When the ground is uneven and sloping
- [ ] C: When weather protection is not fitted
- [ ] D: When the operator is clipped to an anchorage point in the basket

19.12

When providing portable traffic signals on roads used by cyclists and horse riders, what action should you take?

- [ ] A: Locate the signals at bends in the road
- [ ] B: Allow more time for slow-moving traffic by increasing the 'all red' phase of the signals
- [ ] C: Operate the signals manually
- [ ] D: Use 'stop/go' boards only

19.13

Why is it necessary to wear high-visibility clothing when working on roads?

- [ ] A: So road users and plant operators can see you
- [ ] B: So that your supervisor can see you
- [ ] C: Because you were told to
- [ ] D: Because it will keep you warm

19.14

What is the purpose of an on-site risk assessment?

- [ ] A: To ensure there is no risk of traffic build-up due to the works in progress
- [ ] B: To identify hazards and risks in order to ensure a safe system of work
- [ ] C: To ensure that the work can be carried out in reasonable safety
- [ ] D: To protect the employer from prosecution

Answers: 19.11 = B, 19.12 = B 19.13 = A, 19.14 = B

19.15

When working on a dual carriageway with a 60 mph speed limit, unless in the working space, what is the minimum standard of high-visibility clothing must be worn?

- [ ] A: Reflective waistcoat
- [ ] B: Reflective long-sleeved jacket
- [ ] C: None
- [ ] D: Reflective sash

19.16

Why should temporary signing be removed when works are complete?

- [ ] A: To get traffic flowing
- [ ] B: It is a legal requirement
- [ ] C: To allow the road to be opened fully
- [ ] D: To reuse signs on new job

19.17

When should installed signs and guarding equipment be inspected?

- [ ] A: After it has been used
- [ ] B: Once a week
- [ ] C: Before being used
- [ ] D: Regularly and at least once every day

19.18

What two site conditions must prevail so that the minimum traffic management can be used?

- [ ] A: Traffic should be heavy
- [ ] B: Visibility is good
- [ ] C: Double parking will be required
- [ ] D: Rush hour
- [ ] E: Period of low risk

19.19

What traffic management is required when carrying out a maintenance job on a motorway?

- [ ] A: The same as would be required on a single carriageway
- [ ] B: A flashing beacon and a 'keep left/right' sign
- [ ] C: A scheme installed by a registered traffic management contractor
- [ ] D: Ten 1-metre high cones and a 1-metre high 'men working' sign

Answers: 19.15 = B, 19.16 = B, 19.17 = D, 19.18 = B,E, 19.19 = C

19.20

What is the minimum traffic management required when carrying out a short term minor maintenance job in a quiet, low-speed side road?

- [ ] A: A flashing amber beacon and a 'keep left/right' arrow
- [ ] B: The same as required for a road excavation
- [ ] C: Five cones and a blue arrow
- [ ] D: Temporary traffic lights

19.21

Mobile works are being carried out by day. A single vehicle is being used. What must be conspicuously displayed on or at the rear of the vehicle?

- [ ] A: Road narrows (left or right)
- [ ] B: A specific task warning sign (for example, gully cleaning)
- [ ] C: A 'keep left/right' arrow
- [ ] D: A 'roadworks ahead' sign

19.22

Some work activities move along the carriageway, such as sweeping, verge mowing and road lining. What is the maximum distance between the 'roadworks ahead' signs?

- [ ] A: 2 miles
- [ ] B: 1 mile
- [ ] C: Half a mile
- [ ] D: Quarter of a mile

19.23

What action is required when a vehicle fitted with a direction arrow is travelling from site to site?

- [ ] A: Point the direction arrow up
- [ ] B: Travel slowly from site to site
- [ ] C: Point the direction arrow down
- [ ] D: Cover or remove the direction arrow

19.24

Signs placed on footways must be located so that they:

- [ ] A: block the footway
- [ ] B: can be read by site personnel
- [ ] C: do not create a hazard for pedestrians
- [ ] D: can be easily removed

19.25

What should you do if drivers approaching roadworks cannot see the advance signs clearly because of poor visibility or obstructions caused by road features?

- [ ] A: Place additional signs in advance of the works
- [ ] B: Extend the safety zones
- [ ] C: Extend the sideways clearance
- [ ] D: Lengthen the lead-in taper

Answers: 19.20 = A, 19.21 = C, 19.22 = B, 19.23 = D, 19.24 = C, 19.25 = A

19.26

How should a portable traffic light cable that crosses a road be protected?

- [ ] A: It should be secured firmly to the road surface
- [ ] B: A cable crossing protector must be used with 'ramp' warning signs
- [ ] C: It can be unprotected if less than 10 mm diameter
- [ ] D: It should be placed in a slot cut into the road surface

19.27

What action is required where passing traffic may block the view of signs?

- [ ] A: Signs must be larger
- [ ] B: Signs must be duplicated on both sides of the road
- [ ] C: Signs must be placed higher
- [ ] D: Additional signs must be placed in advance of the works

19.28

In which two places would you find information on the distances for setting out the signs in advance of the works under different road conditions?

- [ ] A: In the Traffic Signs Manual (Chapter 8)
- [ ] B: In the 'Pink Book'
- [ ] C: On the back of the sign
- [ ] D: In the specification for highway works
- [ ] E: The Code of Practice ('Red Book')

19.29

Signs, lights and guarding equipment must be properly secured:

- [ ] A: with sacks containing fine granular material set at a low level
- [ ] B: by roping them to concrete blocks or kerb stones
- [ ] C: to prevent them being stolen
- [ ] D: by iron weights suspended from the frame by chains or other strong material

19.30

Which is not an approved means of controlling traffic at roadworks?

- [ ] A: Priority signs
- [ ] B: Police supervision
- [ ] C: Hand signals by operatives
- [ ] D: A give-and-take system

Answers: 19.26 = B, 19.27 = B, 19.28 = A,E, 19.29 = A, 19.30 = C

19.31

What action is required if a vehicle detector on temporary traffic lights becomes defective?

- [ ] A: Control traffic at the defective end using hand signals
- [ ] B: Operate on 'all red' and call the service engineer
- [ ] C: Operate on 'fixed time' or 'manual' and call the service engineer
- [ ] D: Switch the lights off until the supervisor arrives on site

19.32

Portable traffic signals are assembled and placed:

- [ ] A: as speedily as possible
- [ ] B: in an organised manner to a specified sequence
- [ ] C: during the night
- [ ] D: as work starts each morning

19.33

What action is required where it is not possible to maintain the correct safety zone?

- [ ] A: Barrier off the working space
- [ ] B: Place additional advance signing
- [ ] C: Use extra cones on the lead-in taper
- [ ] D: Stop work and consult your supervisor

19.34

In which of the following circumstances can an operative enter the safety zone?

- [ ] A: To store unused plant
- [ ] B: To maintain cones and signs
- [ ] C: To park site vehicles
- [ ] D: To store materials

19.35

On a dual carriageway, a vehicle driven by a member of the public enters the coned-off area. What action do you take?

- [ ] A: Remove a cone and direct the driver back on to the live carriageway
- [ ] B: Ignore them
- [ ] C: Shout and wave them off site
- [ ] D: Assist them to leave the site safely via the nearest designated exit

19.36

You are the driver of a vehicle and lose sight of the signaller while reversing your vehicle. What do you do?

- [ ] A: Continue reversing slowly
- [ ] B: It is OK to continue reversing provided your vehicle is equipped with a klaxon and flashing lights
- [ ] C: Stop and locate the signaller
- [ ] D: Find someone else to watch you reverse safely

Answers: 19.31 = C, 19.32 = B, 19.33 = D, 19.34 = B, 19.35 = D, 19.36 = C

19.37

If you are working after dark, is mobile plant exempt from the requirement to show lights?

- [ ] A: Yes, on all occasions
- [ ] B: Yes if authorised by the supervisor
- [ ] C: Only if they are not fitted to the machine as standard
- [ ] D: Not in any circumstances

19.38

What is the purpose of the 'safety zone'?

- [ ] A: To indicate the works area
- [ ] B: To protect you from the traffic and the traffic from you
- [ ] C: To allow extra working space in an emergency
- [ ] D: To give a safe route around the working area

19.39

When should you switch on the amber flashing beacon fitted to your vehicle?

- [ ] A: At all times
- [ ] B: When travelling to and from the depot
- [ ] C: When the vehicle is being used as a works vehicle
- [ ] D: Only in poor visibility

19.40

You are competent and authorised to drive rollers. Before driving a new type of roller that you haven't driven before, which two of the following must you do?

- [ ] A: Hold a current driving licence
- [ ] B: Drive the roller slowly
- [ ] C: Read the operating instructions
- [ ] D: Familiarise yourself with the controls
- [ ] E: Carry your plant operating certificate

19.41

When driving into a site works access on a motorway what must you do approximately 200 metres before the access?

- [ ] A: Switch on the vehicle hazard lights
- [ ] B: Switch on flashing amber beacon
- [ ] C: Switch on the headlights
- [ ] D: Switch on the flashing amber beacon and the appropriate indicator

Answers: 19.37 = D, 19.38 = B, 19.39 = C, 19.40 = C,D, 19.41 = D

19.42

Lifting equipment for carrying persons, for example a cherry picker, must be thoroughly examined by a competent person every:

- [ ] A: 12 months
- [ ] B: 24 months
- [ ] C: 18 months
- [ ] D: 6 months

19.43

When towing a trailer fitted with independent brakes, what must you do?

- [ ] A: Fit a safety chain
- [ ] B: Fit a cable that applies the trailer's brakes if the tow hitch fails
- [ ] C: Use rope to secure the trailer to the tow hitch
- [ ] D: Drive at a maximum speed of 25 mph

19.44

From a safety point of view, diesel must not be used to prevent asphalt sticking to the bed of lorries because:

- [ ] A: it will create a slipping hazard
- [ ] B: it will corrode the bed of the lorry
- [ ] C: it will create a fire hazard
- [ ] D: it will react with the asphalt, creating explosive fumes

19.45

When getting off plant and vehicles, what must you do?

- [ ] A: Look before you jump
- [ ] B: Use the wheels and tyres for access
- [ ] C: Maintain 3 points of contact with the vehicle
- [ ] D: Jump down facing the vehicle

19.46

When leaving plant unattended what should you do?

- [ ] A: Leave amber flashing beacon on
- [ ] B: Apply brake, switch off engine, remove the key
- [ ] C: Leave in a safe place with the engine ticking over
- [ ] D: Park with blocks under the front wheels

19.47

What must you have before towing a compressor on the highway?

- [ ] A: Permission from your supervisor
- [ ] B: The correct class of driving licence
- [ ] C: Permission from the police
- [ ] D: Permission from the compressor hire company

Answers: 19.42 = D, 19.43 = B, 19.44 = A, 19.45 = C, 19.46 = B, 19.47 = B

19.48

Who is responsible for the security of the load on a vehicle?

- [ ] A: The driver's supervisor
- [ ] B: Police
- [ ] C: The driver
- [ ] D: The driver's company

19.49

What should be used to protect the public from a shallow excavation in a public footway?

- [ ] A: Pins and bunting
- [ ] B: Nothing
- [ ] C: Cones
- [ ] D: Barriers with tapping rails

19.50

When materials have to be kept on site overnight, what should you do?

- [ ] A: Don't stack them above 2 metres high
- [ ] B: Stack the materials safely and in a secure area
- [ ] C: Put pins and bunting around them
- [ ] D: Only stack them on the grass verge

Answers: 19.48 = C, 19.49 = D, 19.50 = B

20.1

If working from a cherry-picker, you should attach your safety lanyard to:

- ☐ A: a strong part of the structure you are working on
- ☐ B: a secure anchorage point inside the platform
- ☐ C: a secure point on the boom of the machine
- ☐ D: a scaffold guard-rail

20.2

If a working platform is 4 metres above the ground, the foot of the access ladder should be placed:

- ☐ A: 1 metre out
- ☐ B: 2 metres out
- ☐ C: 3 metres out
- ☐ D: 4 metres out

20.3

What should be included in a safety method statement for working at height? Give three answers.

- ☐ A: The cost of the job and time it will take
- ☐ B: The sequence of operations and the equipment to be used
- ☐ C: How much insurance cover will be required
- ☐ D: How falls are to be prevented
- ☐ E: Who will supervise the job on site

20.4

Scaffold towers may be erected by:

- ☐ A: anyone who has the instruction book
- ☐ B: anyone who is competent and authorised
- ☐ C: advanced scaffolders only
- ☐ D: an employee of the hire company only

20.5

When working above public areas, your first consideration should be:

- ☐ A: minimise the number of people below at any one time
- ☐ B: to prevent complaints from the public
- ☐ C: to let the public know what you are doing
- ☐ D: to prevent anything falling on to people below

20.6

A competent person must routinely inspect a scaffold:

- ☐ A: after it is erected and at intervals not exceeding 7 days
- ☐ B: only after it has been erected
- ☐ C: after it is erected and then at monthly intervals
- ☐ D: after it is erected and then at intervals not exceeding 10 days

Answers: 20.1 = B, 20.2 = A, 20.3 = B,D,E, 20.4 = B, 20.5 = D, 20.6 = A

20.7

When covering rooflights, what two requirements should the covers meet?

- [ ] A: They are made from the same material as the roof covering
- [ ] B: They are made from clear material to allow the light through
- [ ] C: They are strong enough to take the weight of any load placed on it
- [ ] D: They are waterproof and windproof
- [ ] E: They are fixed in position to stop them being dislodged

20.8

You are working at height from a cherry picker when the weather becomes very windy. Your first consideration should be:

- [ ] A: to tie all lightweight objects to the hand-rails of the basket
- [ ] B: to decide whether the machine will remain stable
- [ ] C: to tie the cherry picker basket to the structure you are working on
- [ ] D: to clip your lanyard to the structure that you are working on

20.9

What is the main reason for using a safety net or inflatable air bags rather than harness and lanyard?

- [ ] A: Safety nets or air bags are cheaper to use
- [ ] B: A safety harness can be uncomfortable to wear
- [ ] C: Harnesses and lanyards have to be inspected
- [ ] D: Safety nets or air bags are collective fall arrest measures

20.10

You are working above a safety net. You notice the net is damaged. What should you do?

- [ ] A: Work somewhere away from the damaged area of net
- [ ] B: Stop work and report it
- [ ] C: Tie the damaged edges together using the net test cords
- [ ] D: See if you can get hold of a harness and lanyard

Answers: 20.7 = C,E, 20.8 = B, 20.9 = D, 20.10 = B

20.11

What is the main reason for not allowing debris to gather in safety nets?

- A: It will overload the net
- B: It looks untidy from below
- C: It could injure someone who falls into the net
- D: Small pieces of debris may fall through the net

20.12

You are working at height, but the securing cord for a safety net is in your way. What should you do?

- A: Untie the cord, carry out your work and tie it up again
- B: Untie the cord, but ask the net riggers to re-tie it when you have finished
- C: Tell the net riggers that you are going to untie the cord
- D: Leave the cord alone and report the problem

20.13

Ideally, a safety net should be rigged:

- A: immediately below where you are working
- B: 2 metres below where you are working
- C: 6 metres below where you are working
- D: at any height below the working position

20.14

If a safety lanyard has damaged stitching, you should:

- A: use the lanyard if the damaged stitching is less than 2 inches long
- B: get a replacement lanyard
- C: do not use the damaged lanyard and work without one
- D: use the lanyard if the damaged stitching is less than 6 inches long

20.15

What is the MAIN danger of leaving someone who has fallen suspended in a harness for too long?

- A: They will become bored
- B: They may try to climb back up the structure and fall again
- C: They may suffer severe discomfort and lose consciousness
- D: It is a distraction for other workers

20.16

What is the recommended maximum height for a free-standing mobile tower when used indoors?

- A: There is no height restriction
- B: Three lifts
- C: As specified by the manufacturer
- D: Three times the longest base dimension

Answers: 20.11 = C, 20.12 = D, 20.13 = A, 20.14 = B, 20.15 = C, 20.16 = C

20.17

After gaining access to the platform of a mobile tower, the first thing you should do is:

- A: check that the tower's brakes are locked on
- B: check that the tower has been correctly assembled
- C: close the access hatch to stop people or equipment from falling
- D: check that the tower does not rock or wobble

20.18

Before a mobile tower is moved, you must first:

- A: clear the platform of people and equipment
- B: get a Permit to Work
- C: get approval from the Principal Contractor
- D: make arrangements with the forklift truck driver

20.19

Which of these must happen before any roof work starts?

- A: A risk assessment must be carried out
- B: The operatives working on the roof must be trained in the use of safety harnesses
- C: Permits to Work must be issued to those allowed to work on the roof
- D: A weather forecast must be obtained

20.20

Edge protection is designed to:

- A: make access to the roof easier
- B: secure tools and materials close to the edge
- C: stop rainwater running off the roof onto workers below
- D: prevent people and materials falling

Answers: 20.17 = C, 20.18 = A, 20.19 = A, 20.20 = D

20.21

Before climbing a ladder you notice that it has a rung missing near the top. What should you do?

- A: Do not use the ladder and immediately report the defect
- B: Use the ladder but take care when stepping over the position of the missing rung
- C: Turn the ladder over so that the missing rung is near the bottom and use it
- D: See if you can find a piece of wood to replace the rung

20.22

When can someone who is not a scaffolder remove parts of a scaffold?

- A: If the scaffold is not more than 2 lifts in height
- B: As long as a scaffolder refits the parts after the work has finished
- C: Never, only competent scaffolders can remove the parts
- D: Only if it is a tube and fittings scaffold

20.23

During your work, you find that a scaffold tie is in your way. What should you do?

- A: Ask a scaffolder to remove it
- B: Remove it yourself and then replace it
- C: Remove it yourself but get a scaffolder to replace it
- D: Report the problem to your supervisor

20.24

How far should a ladder extend above the stepping-off point?

- A: 3 rungs
- B: 2 rungs
- C: 1 metre
- D: Half a metre

20.25

When using ladders for access, what is the maximum vertical distance between landings?

- A: 5 metres
- B: There is no maximum
- C: 9 metres
- D: 30 metres

Answers: 20.21 = A, 20.22 = C, 20.23 = D, 20.24 = C, 20.25 = C

20.26

On a working platform, the maximum permitted gap between the guard-rails is:

- [ ] A: 350 mm
- [ ] B: 470 mm
- [ ] C: 490 mm
- [ ] D: 510 mm

20.27

When should guard-rails be fitted to a working platform?

- [ ] A: If it is possible to fall 2 metres
- [ ] B: At any height if a fall could result in an injury
- [ ] C: If it is possible to fall 3 metres
- [ ] D: Only if materials are being stored on the working platform

20.28

When are you allowed to carry materials up a ladder on to a scaffold platform? Give two answers.

- [ ] A: If it is a Class 3 ladder
- [ ] B: If you can keep one hand on the ladder at all times
- [ ] C: If you are only going up to the first lift
- [ ] D: If it is a system-built scaffold
- [ ] E: If the load and conditions are such that you are stable at all times

20.29

You need to store materials on a flat roof that cannot be fitted with edge protection. What three things must you ensure?

- [ ] A: The materials are stored in such a way that they cannot fall
- [ ] B: A cantilever safety net is installed below the roof edge
- [ ] C: The stored materials do not endanger others working on the roof
- [ ] D: The area below the roof is kept clear at all times
- [ ] E: There is safe access to the stored materials

20.30

The Beaufort Scale is important when working at height because it measures:

- [ ] A: air temperature
- [ ] B: the load-bearing capacity of a flat roof
- [ ] C: wind speed
- [ ] D: the load-bearing capacity of a scaffold

Answers: 20.26 = B, 20.27 = B, 20.28 = B,E, 20.29 = A,C,E, 20.30 = C

20.31

Before starting work at height, the weather forecast says the wind will increase to 'Force 7'. What does this mean?

- [ ] A: A moderate breeze that can raise light objects, such as dust and leaves
- [ ] B: A near gale that will make it difficult to move about and handle materials
- [ ] C: A gentle breeze that you can feel on your face
- [ ] D: Hurricane winds that will uproot trees and cause structural damage

20.32

You are on a cherry picker, but it does not quite reach where you need to work. What should you do?

- [ ] A: Use a step-ladder balanced on the machine platform
- [ ] B: Extend the machine fully and stand on the guard-rails
- [ ] C: Abandon the machine and use a long extending ladder
- [ ] D: Do not carry out the job until you have an alternative means of access

20.33

If you are working at height and operating a Mobile Elevating Work Platform, when is it acceptable for someone to use the ground-level controls?

- [ ] A: If the person on the ground is trained and you are not
- [ ] B: In an emergency only
- [ ] C: If you need to jump off the Mobile Elevating Work Platform to gain access to the work
- [ ] D: If you need both hands free to carry out the job

20.34

A 'Class 3' ladder is:

- [ ] A: for domestic use only and must not be used on site
- [ ] B: of industrial quality and can be used safely
- [ ] C: a ladder that has been made to a European Standard
- [ ] D: made of insulating material and can be used near to overhead cables

Answers: 20.31 = B, 20.32 = D, 20.33 = B, 20.34 = A

20.35

When is it acceptable to jump off a Mobile Elevating Work Platform on to a high level work platform?

- [ ] A: If the work platform is fitted with edge protection
- [ ] B: If the machine operator stays in the basket
- [ ] C: Not under any circumstances
- [ ] D: If the machine is being operated from the ground-level controls

20.36

How will you know the maximum weight or number of people that can be lifted safely on a mobile elevating work platform?

- [ ] A: The weight limit is reached when the platform is full
- [ ] B: It will say on the Health and Safety Law poster
- [ ] C: You will be told during site induction
- [ ] D: From an information plate fixed to the machine

20.37

When working at height, which of these is the safest way to transfer waste materials to ground level?

- [ ] A: Through a waste chute directly into a skip
- [ ] B: Ask someone below to keep the area clear of people then throw the waste down
- [ ] C: Erect barriers around the area where the waste will land
- [ ] D: Bag or bundle up the waste before throwing it down

20.38

You need to use a ladder to access a roof. The only place to rest the ladder is on a run of plastic gutter. What two things should you consider doing?

- [ ] A: Rest the ladder on a gutter support bracket
- [ ] B: Rest the ladder against the gutter, climb it and quickly tie it off
- [ ] C: Find another way to access the roof
- [ ] D: Use a proprietary 'stand-off' device that allows the ladder to rest against the wall
- [ ] E: Position the ladder at a shallow angle so that it rests below the gutter

Answers: 20.35 = C, 20.36 = D, 20.37 = A, 20.38 = C,D

20.39

If using inflatable air bags as a means of fall arrest, the inflation pump must:

- [ ] A: be electrically-powered
- [ ] B: be switched off from time to time to avoid over-inflation
- [ ] C: run all the time while work is carried out at height
- [ ] D: be switched off when the air bags are full

20.40

Why is it dangerous to use inflatable air bags that are too big for the area to be protected?

- [ ] A: They will exert a sideways pressure on anything that is containing them
- [ ] B: The pressure in the bags will cause them to burst
- [ ] C: The inflation pump will become overloaded
- [ ] D: They will not fully inflate

20.41

When is it most appropriate to use a safety harness and lanyard for working at height?

- [ ] A: Only when the roof has a steep pitch
- [ ] B: Only when crossing a flat roof with clear roof-lights
- [ ] C: Only when all other options for fall prevention have been ruled out
- [ ] D: Only when materials are stored at height

20.42

When is it safe to use a scissor lift on soft ground?

- [ ] A: When the ground is dry
- [ ] B: When the machine can stand on scaffold planks laid over the soft ground
- [ ] C: When stabilisers or outriggers can be deployed onto solid ground
- [ ] D: Never

20.43

Who should install safety nets?

- [ ] A: Anyone who knows how to tie the correct knots
- [ ] B: Anyone who will be working above the net
- [ ] C: Any competent and authorised person
- [ ] D: Any steel erector

Answers: 20.39 = C, 20.40 = A, 20.41 = C, 20.42 = C, 20.43 = C

20.44

You need to cross a roof. How do you establish if it is fragile?

- [ ] A: Tread gently and listen for cracking
- [ ] B: Look at the risk assessment or method statement
- [ ] C: Look at the roof surface and make your own assessment
- [ ] D: It does not matter if you walk along a line of bolts

20.45

After gaining access to a roof, you notice some overhead cables within reach. What should you do?

- [ ] A: Keep away from them while you work but remember they are there
- [ ] B: Confirm that it is safe for you to be on the roof
- [ ] C: Make sure that you are using a wooden ladder
- [ ] D: Hang coloured bunting from them to remind you they are there

20.46

You have to lean over an exposed edge while working at height. How should you wear your safety helmet?

- [ ] A: Titled back on your head so that it doesn't fall off
- [ ] B: Take your helmet off while leaning over then put it on again afterwards
- [ ] C: Wear the helmet as usual but use the chin-strap
- [ ] D: Wear the helmet back to front whilst leaning over

20.47

When trying to clip your lanyard to an anchor point, you find the locking device does not work. What should you do?

- [ ] A: Carry on working and report it later
- [ ] B: Tie the lanyard in a knot round the anchor
- [ ] C: Stop work and report to your supervisor
- [ ] D: Just carry on working without it

Answers: 20.44 = B, 20.45 = B, 20.46 = C, 20.47 = C

20.48

An outdoor tower scaffold has stood overnight in high winds and heavy rain. What should you consider before you use the scaffold?

- [ ] A: That the brakes still work
- [ ] B: Tying the scaffold to the adjacent structure
- [ ] C: That the scaffold is inspected by a competent person
- [ ] D: That the platform hatch still works correctly

20.49

What is the main disadvantage of an aluminium scaffold tower?

- [ ] A: The aluminium will corrode in wet weather
- [ ] B: It cannot be built more than 2 lifts high
- [ ] C: Materials cannot be stored on the working platform
- [ ] D: The lack of weight means it can be displaced by high winds

20.50

Which type of scaffold tie can be removed by someone who is not a scaffolder?

- [ ] A: A box tie
- [ ] B: A ring tie
- [ ] C: A reveal tie
- [ ] D: None

Answers: 20.48 = C, 20.49 = D, 20.50 = D

21 Lifts and Escalators

21.1

A gap of what size or more, between the edge of the work platform and the hoistway wall, is regarded as a fall hazard?

- A: 250mm
- B: 300mm
- C: 330mm
- D: 450mm

21.2

Before entering the pit of an operating lift, you must first:

- A: fit pit props
- B: verify the pit stop switch
- C: switch the lift off
- D: position the access ladder

21.3

When is it acceptable to work on the top of a car without a top-of-car control station?

- A: When the unit has been locked and tagged out
- B: When there are two engineers
- C: When there is no other way
- D: When single-man working

21.4

Who should fit a padlock and tag to the lock-out guard?

- A: Anyone working on the unit
- B: The engineer who fitted the lock-out device
- C: The senior engineer
- D: Anyone

21.5

What is the frequency of the statutory period of inspection for lifting equipment used to support people?

- A: At least monthly
- B: At least every 6 months
- C: At least every 12 months
- D: Once every 2 years

21.6

Which statement is incorrect?

- A: A stop switch must be within 15m of the front of the car
- B: The car top should be clean and free from grease and oil spills
- C: You should secure your tools out of your standing area when working on top of the car
- D: Before trying to gain access to the hoistway, you should decide whether the work will need the power supply to be live

Answers: 21.1 = B, 21.2 = B, 21.3 = A, 21.4 = A, 21.5 = B, 21.6 = A

21.7

When working on an energised car, which statement is true?

- [ ] A: Always attach your lanyard to the car top while standing on the landing
- [ ] B: Make sure you step onto the landing with your lanyard still attached to the car
- [ ] C: Always check the lanyard is unclipped before getting off the car top
- [ ] D: Ensure that your lanyard is clipped to a guide bracket or similar anchorage on the shaft

21.8

You arrive on site and find the lift mains isolator is switched off. What do you do?

- [ ] A: Switch it on and get on with your work
- [ ] B: Switch it on and check the safety circuits to see if there is a fault
- [ ] C: Contact the person in control of the premises to find out if they had switched the lift off or if anyone else is working on site
- [ ] D: Shout down the shaft and if no one responds switch it on and get on with your work

21.9

If a counterweight screen is not fitted or has been removed, what should you do before starting work?

- [ ] A: Carry out a further risk assessment to establish a safe system of work
- [ ] B: Nothing – just get on with the job as normal
- [ ] C: Give a tool box talk on guarding
- [ ] D: Issue and wear appropriate Personal Protective Equipment

21.10

What is the last thing you should do before alighting from a car top through open landing doors when the car-top control is within 1 metre of the landing threshold?

- [ ] A: Set the car-top control to 'Test'
- [ ] B: Ensure that the car-top stop button is set to 'Stop' and the car-top control remains set to 'Test'
- [ ] C: Turn off the shaft lights and switch the car-top control to 'Normal' operation
- [ ] D: Press the stop button and switch the car-top control to 'Normal' operation

Answers: 21.7 = C, 21.8 = C, 21.9 = A, 21.10 = B

21.11

If the landing doors are to be open while work goes on in the lift pit, what precaution must you take?

- [ ] A: Erect a barrier and secure it in front of the landing doors
- [ ] B: Post a notice on the wall next to the doors
- [ ] C: Do the job when there are few people about
- [ ] D: Ask someone to guard the open doors while you work

21.12

It is essential that an authorised person working alone does which two of the following?

- [ ] A: Before starting work, registers their presence with the site representative
- [ ] B: Ensures their time sheet is accurate and countersigned
- [ ] C: Establishes suitable arrangements to ensure the monitoring of their well being
- [ ] D: Notifies the site manager of the details of the work
- [ ] E: Ensures that the lift pit is free from water and debris

21.13

Before accessing the car top of an operating lift, what is the first thing to do after opening the landing doors?

- [ ] A: Make sure the lift has stopped
- [ ] B: Press the car-top stop button
- [ ] C: Chock the landing doors open
- [ ] D: Put the car-top control in the 'Test' position

21.14

Which of the following types of fire extinguisher should not be used if a fire occurs in a lift or escalator controller?

- [ ] A: Halon
- [ ] B: Water
- [ ] C: Dry powder
- [ ] D: Carbon dioxide

21.15

When you gain access to a car top, you should test that the car-top stop switch operates correctly by:

- [ ] A: trying to move the car in the up direction
- [ ] B: trying to move the car in the down direction
- [ ] C: measuring with a multimeter
- [ ] D: flicking the switch on and off rapidly

Answers: 21.11 = A, 21.12 = A,C, 21.13 = A, 21.14 = B, 21.15 = B

21.16

When using authorised lifting tackle marked with its safe working load, which statement is true?

- [ ] A: Never exceed the safe working load
- [ ] B: The safe working load is only guidance
- [ ] C: Halve the safe working load if the equipment is damaged
- [ ] D: Double the safe working load if people are to be lifted

21.17

What checks do you need to carry out before using lifting equipment?

- [ ] A: Carry out a drop check
- [ ] B: Check that it is free from defects and has a current inspection certificate
- [ ] C: Check to ensure the chains are knotted to correct length
- [ ] D: Date of manufacture of the lifting tackle

21.18

What should you do if the lifting tackle you are about to use is defective?

- [ ] A: Only use it for half its safe working load
- [ ] B: Only use it for small lifts under 1 tonne
- [ ] C: Do not use it and inform your supervisor
- [ ] D: Try to fix it

21.19

Who is permitted to undertake the safe release of trapped passengers?

- [ ] A: The site foreman
- [ ] B: Only a trained and authorised person
- [ ] C: Anyone
- [ ] D: Only the emergency services

21.20

If landing doors are not fitted to a lift on a construction site, what is the minimum height of the barrier that must be fitted instead?

- [ ] A: 650 mm
- [ ] B: 740 mm
- [ ] C: 810 mm
- [ ] D: 950 mm

Answers: 21.16 = A, 21.17 = B, 21.18 = C, 21.19 = B, 21.20 = D

Lifts and Escalators

21.21

What is the main cause of injury and absence to employees within the lift and escalator industry?

- [ ] A: Falls
- [ ] B: Electrocution
- [ ] C: Contact with moving parts
- [ ] D: Manual handling

21.22

A set of chain blocks has been delivered to site with a certificate stating they were inspected by a competent person a month before. The hook is obviously damaged. What action do you take?

- [ ] A: Use the blocks as the certificate is in date
- [ ] B: Do not use the blocks and inform your supervisor
- [ ] C: Use the blocks at half the safe working load
- [ ] D: Use the blocks until replacement equipment arrives

21.23

Following the initial inspection, how often should a scaffold in a lift shaft be inspected by a competent person?

- [ ] A: At least every day
- [ ] B: At least every 7 days
- [ ] C: At least every 14 days
- [ ] D: There is no set period between inspections

21.24

What should be fitted to the main sheave and divertor to prevent injury from rotating equipment?

- [ ] A: Movement sensors
- [ ] B: Guards
- [ ] C: Clutching assembly
- [ ] D: Safety notices

21.25

When working on electrical lift-control equipment, what are the appropriate tools and equipment?

- [ ] A: Insulated tools and an insulating mat
- [ ] B: Non-insulated tools
- [ ] C: Any tools and an insulating mat
- [ ] D: No tools are allowed near electrical equipment

21.26

When installing a partially enclosed or observation lift, what safe system of work can you use to prevent injury to people below?

- [ ] A: Put up a sign
- [ ] B: Do not use heavy tools
- [ ] C: Secure tools to prevent them falling off
- [ ] D: Only carry out essential work using minimum tools

Answers: 21.21 = D, 21.22 = B, 21.23 = B, 21.24 = B, 21.25 = A, 21.26 = C

21.27

What method of storing oil cans or drums must be used to prevent any leakage?

- [ ] A: One that is easy to identify and remove if there is any leak or damage
- [ ] B: On their sides and chocked to prevent movement
- [ ] C: Upside down to prevent water penetrating the screw top
- [ ] D: In a bunded enclosure

21.28

What needs to be checked before any welding or cutting takes place in the lift installation?

- [ ] A: The weight and size of the welding equipment
- [ ] B: How long the task will take
- [ ] C: If a 'hot work' permit or permit to work is required
- [ ] D: If the local fire services require notifying

21.29

Rings, bracelets, wrist watches, necklaces etc must not be worn:

- [ ] A: when working near or on electrical or moving equipment
- [ ] B: when working on site generally
- [ ] C: when driving a company vehicle
- [ ] D: after leaving home for work

21.30

When using arc-welding equipment you must take precautions. Which two of the following are potential hazards when using this equipment?

- [ ] A: Noise
- [ ] B: Fire from hot spent electrodes
- [ ] C: Arc flash to people working close to the welding point
- [ ] D: Incompatible materials
- [ ] E: Extended exposure to infrared light

21.31

If the trap-door or hatch has to be left open while you work in the machine room, what must you ensure?

- [ ] A: That a sign is posted to warn others
- [ ] B: That the distance from the trapdoor/hatch to floor below does not exceed two metres
- [ ] C: That there is sufficient light available for work
- [ ] D: That a barrier is put in place

21.32

To prevent unauthorised access to unoccupied machine equipment space, what must you ensure?

- [ ] A: That the access door is locked
- [ ] B: That a sign is posted
- [ ] C: That the power supply is isolated
- [ ] D: That a person is posted to prevent access

Answers: 21.27 = D, 21.28 = C, 21.29 = A, 21.30 = B,C, 21.31 = D, 21.32 = A

21.33

When working in the pit, the lift should be positioned towards the top of the shaft unless:

- [ ] A: the hydraulic fluid level is low
- [ ] B: work needs to be done on the underside of the lift
- [ ] C: you are testing the buffers
- [ ] D: the power supply is cut

21.34

What should be applied to the main isolator of a traction lift to prevent accidental starting?

- [ ] A: A warning notice
- [ ] B: Lock-out device
- [ ] C: Residual current device
- [ ] D: Lower-rated fuses

21.35

What is the correct method of disposal for used or contaminated oil?

- [ ] A: Decant it into a sealed container and place in a skip
- [ ] B: Through a registered waste process
- [ ] C: Dilute it with water and pour down a sink
- [ ] D: Pour it down a roadside drain

21.36

Who is responsible for the keys when a padlock has been applied to a lock-out device?

- [ ] A: The individual applying the lock
- [ ] B: Supervisor
- [ ] C: Manager
- [ ] D: The person nearest the lock-out device

21.37

If work is to be done on electrical lift equipment, and the main isolator does not have a lock-out device, what is an alternative method of isolating the supply?

- [ ] A: Detach the main power supply
- [ ] B: Place insulation material in the contactors
- [ ] C: There is no alternative to a lock-out device for isolating the supply
- [ ] D: Withdraw and retain the fuse and fix notices warning that the machinery is being worked on

Answers: 21.33 = B, 21.34 = B, 21.35 = B, 21.36 = A, 21.37 = D

21.38

The main contractor wants to use the unfinished lift to move some equipment to an upper floor. What should you do?

- [ ] A: Help to ensure the load is correctly positioned
- [ ] B: Refer him to your supervisor
- [ ] C: Ask for the weight of the equipment
- [ ] D: Allow him to use the lift but take no responsibility

21.39

While installing a new rope, you notice a damaged section where something heavy has fallen onto the coil. Do you:

- [ ] A: fit the rope anyway?
- [ ] B: cut out the damaged section?
- [ ] C: reject all the ropes?
- [ ] D: add an extra termination?

21.40

A large heavy balance weight frame is delivered to site on a lorry with no crane. There is no lifting equipment available on site. Do you:

- [ ] A: unload it manually?
- [ ] B: arrange for it to be re-delivered on a suitable lorry?
- [ ] C: slide it down planks?
- [ ] D: tip the load off?

21.41

What is required on each landing of a new lift shaft before entrances and doors are fitted?

- [ ] A: A warning notice
- [ ] B: Substantial secure barrier to prevent falls
- [ ] C: Orange plastic netting across the opening
- [ ] D: Lighting

21.42

A lifting beam at the top of the lift shaft is marked with a SWL of 800kg but the brickwork around the beam is cracked and appears to be loose. Do you:

- [ ] A: use the beam as normal?
- [ ] B: only lift loads not exceeding 400 kg?
- [ ] C: not use the beam and speak to your supervisor?
- [ ] D: de-rate the beam by 75%?

21.43

To prevent injury from an overspeed governor, what is fitted?

- [ ] A: A rope
- [ ] B: A restrictor
- [ ] C: A guard
- [ ] D: A switch

Answers: 21.38 = B, 21.39 = C, 21.40 = B, 21.41 = B, 21.42 = C, 21.43 = C

21.44

When handling stainless steel car panels, which of the following items of Personal Protective Equipment should you wear in addition to safety footwear?

- [ ] A: Rigger gloves
- [ ] B: Barrier cream
- [ ] C: PVC gloves
- [ ] D: Hearing protection

21.45

At what stage in the installation of a lift should guarding be fitted to the lift machine?

- [ ] A: At the end of the job
- [ ] B: During commissioning
- [ ] C: When handing over to the client
- [ ] D: Before the machine can be operated

21.46

Before gaining access into the escalator or passenger conveyor, it is essential that:

- [ ] A: the mains switch is locked and tagged out
- [ ] B: the mains switch is in the 'On' position
- [ ] C: all steps are removed
- [ ] D: the drive mechanism is lubricated

21.47

What is secured at the entry/exit points of an escalator/passenger conveyor to prevent people falling into the machine or machine space?

- [ ] A: Safety barriers
- [ ] B: Safety notices
- [ ] C: Escalator machine equipment guards
- [ ] D: Machine tank covers

21.48

What must you do before moving the steps or pallet band of an escalator or passenger conveyor?

- [ ] A: Check there are no sharp edges on the steps
- [ ] B: Check there is a clear route of escape
- [ ] C: Check that no unauthorised people are on the equipment
- [ ] D: Check that a fire extinguisher is available

Answers: 21.44 = A, 21.45 = D, 21.46 = A, 21.47 = A, 21.48 = C

21.49

If the escalator or passenger conveyor has an external machine room, access doors should be:

- [ ] A: capable of being locked and be marked with the appropriate safety sign
- [ ] B: smoke proof in case of a fire
- [ ] C: unlocked at all times in case of an emergency
- [ ] D: capable of being locked on the inside only and be marked with the appropriate safety sign

Answer: 21.49 = A

Heating, Ventilating, Air Conditioning and Refrigeration (HVACR)

If you are taking one if the HVACR tests, you will be asked questions from sections 1-15 plus questions on one of the following subjects:

- Section 22 HVACR: Domestic Heating and Plumbing Services
- Section 23 HVACR: Pipefitting/welding (industrial and commercial)
- Section 24 HVACR: Ductwork
- Section 25 HVACR: Refrigeration and Air Conditioning
- Section 26 HVACR: Services and Facilities

HVACR: Domestic Heating and Plumbing Services

22.1

If you find a coloured wire sticking out of an electrical plug, which is the correct action to take?

- [ ] A: Push it back into the plug and carry on working
- [ ] B: Pull the wire clear of the plug and report it to your supervisor
- [ ] C: Mark the item as defective and follow your company procedure for defective items
- [ ] D: Take the plug apart and carry out a repair

22.2

Extension leads in use on a site should be:

- [ ] A: located so as to prevent a tripping hazard
- [ ] B: laid out in the shortest, most convenient route
- [ ] C: coiled on a drum or cable tidy
- [ ] D: raised on bricks

22.3

What should you do if you need additional temporary wiring for your power tools whilst working on site?

- [ ] A: Find some cable and extend the wiring yourself
- [ ] B: Stop work until an authorised supply has been installed
- [ ] C: Speak to an electrician and ask him to do the temporary wiring
- [ ] D: Disconnect a longer cable serving somewhere else and reconnect it to where you need

22.4

When planning a lifting operation the sequence of operations to enable a lift to be carried out safely should be confirmed in:

- [ ] A: verbal instructions
- [ ] B: a method statement
- [ ] C: a radio telephone message
- [ ] D: a notice in the canteen

22.5

The safe working load (SWL) of lifting equipment is:

- [ ] A: never marked on the equipment but kept with the test certificates
- [ ] B: provided for guidance only
- [ ] C: may be exceeded by no more than 25%
- [ ] D: the absolute maximum safe working load

Answers: 22.1 = C, 22.2 = A, 22.3 = B, 22.4 = B, 22.5 = D

22.6

Before using a ladder you must make sure that:

- A: it is secured to prevent it from moving sideways or sliding outwards
- B: no one else has booked the ladder for their work
- C: an apprentice or workmate is standing by in case you slip and fall
- D: the weather forecast is for a bright, clear day

22.7

When positioning and erecting a stepladder, which of the following is essential for its safe use?

- A: It has a tool tray towards the top of the steps
- B: The restraint mechanism is spread to its full extent
- C: You will be able to reach the job by standing on the top step
- D: A competent person has positioned and erected the steps

22.8

Generally, how many working platforms should be in use at any one time on a mobile tower?

- A: One
- B: Two
- C: Three
- D: Four

22.9

What is the recommended maximum height for a free-standing mobile tower?

- A: There is no restriction
- B: 2 metres
- C: In accordance with the manufacturer's recommendations
- D: 12 metres

22.10

What is the first thing you should do after getting on to the platform of a mobile tower?

- A: Check that the brakes are locked on
- B: Check the mobile tower to make sure that it has been correctly assembled
- C: Close the access hatch to prevent falls of personnel, tools or equipment
- D: Make sure that the tower does not rock or wobble

22.11

What should be done before a mobile tower is moved?

- A: All people and equipment must be removed from the platform
- B: A Permit to Work is required
- C: The Principal Contractor must give their approval
- D: Arrangements must be made with the forklift truck driver

Answers: 22.6 = A, 22.7 = B, 22.8 = A, 22.9 = C, 22.10 = C, 22.11 = A

22.12

What must be done first before any roof work is carried out?

- [ ] A: A risk assessment must be carried out
- [ ] B: The operatives working on the roof must be trained in the use of safety harnesses
- [ ] C: Permits to Work must be issued only to those allowed to work on the roof
- [ ] D: A weather forecast must be obtained

22.13

What is edge protection designed to do?

- [ ] A: Make access to the roof easier
- [ ] B: Secure tools and materials close to the edge
- [ ] C: Prevent rainwater running off the roof onto workers below
- [ ] D: Prevent the fall of people and materials

22.14

Where should Liquefied Petroleum Gas (LPG) cylinders be positioned when supplying an appliance in a site cabin?

- [ ] A: Inside the site cabin in a locked cupboard
- [ ] B: Under the cabin
- [ ] C: Inside the cabin next to the appliance
- [ ] D: Outside the cabin

22.15

If you are working where welding is being carried out, what should be provided to protect you from 'welding flash'?

- [ ] A: Fire extinguishers
- [ ] B: Warning notices
- [ ] C: Screens
- [ ] D: High visibility vest

22.16

How should you position the exhaust of an engine-driven generator which has to be run inside a building?

- [ ] A: Position the exhaust outside the building
- [ ] B: Position the exhaust in a stairwell
- [ ] C: Hang the exhaust in another room
- [ ] D: Position the exhaust in a riser

Answers: 22.12 = A, 22.13 = D, 22.14 = D, 22.15 = C, 22.16 = A

22.17

How should cylinders containing Liquefied Petroleum Gas (LPG) be stored on site?

- A: In a locked cellar with clear warning signs
- B: In a locked external compound at least 3 metres from any oxygen cylinders
- C: As close to the point of use as possible
- D: Covered by a tarpaulin to shield the compressed cylinder from sunlight

22.18

When working in a riser, how should access be controlled?

- A: By a site security operative
- B: By those who are working in it
- C: By the main contractor
- D: By a Permit to Work system

22.19

Which of the following electrical equipment does NOT require portable appliance testing?

- A: Battery-powered rechargeable drill
- B: 110 volt electrical drill
- C: 110 volt portable halogen light
- D: Electric kettle

22.20

What MUST be clearly marked on all lifting equipment?

- A: Name of the manufacturer
- B: The safe working load
- C: Next test date
- D: Specification of material from which made

22.21

What is the first thing you should do after getting on to the platform of a correctly erected mobile tower?

- A: Check that the brakes are locked on
- B: Check the mobile tower to make sure that it has been correctly assembled
- C: Close the access hatch to prevent falls of personnel, tools or equipment
- D: Make sure that the tower does not rock or wobble

22.22

When assembling a mobile tower what major overhead hazard must you be aware of?

- A: Water pipes
- B: Cable trays
- C: False ceilings
- D: Suspended electric cables

Answers: 22.17 = B, 22.18 = D, 22.19 = A, 22.20 = B, 22.21 = C, 22.22 = D

22.23

You spill some oil on the floor and you do not have any absorbent material to clean the area. What should you do?

- [ ] A: Spread it about to lessen the depth
- [ ] B: Keep people out of the area and inform your supervisor
- [ ] C: Do nothing, it will eventually soak into the floor
- [ ] D: Warn other people as they tread through it

22.24

What should folding stepladders be used for?

- [ ] A: General access on site
- [ ] B: Short-term activities lasting minutes that don't involve stretching
- [ ] C: All site activities where a straight ladder cannot be used
- [ ] D: Getting on and off mobile towers

22.25

When a new piece of plant has been installed but has not been commissioned, how should it be left?

- [ ] A: With any valves and switches turned off
- [ ] B: With any valves and switches clearly labelled
- [ ] C: With all valves and switches 'locked off'
- [ ] D: With any valves and switches turned on and ready to use

22.26

If you find a dangerous gas fitting that is likely to cause a death or major injury, to whom MUST a formal report be sent?

- [ ] A: The client
- [ ] B: The gas board
- [ ] C: The Health and Safety Manager
- [ ] D: The Health and Safety Executive

22.27

Who can change a gas valve on a gas boiler?

- [ ] A: A skilled engineer
- [ ] B: A pipefitter
- [ ] C: A competent, CORGI-registered engineer
- [ ] D: Anybody

Answers: 22.23 = B, 22.24 = B, 22.25 = C, 22.26 = D, 22.27 = C

22.28

Who can solder a fitting on an isolated copper gas pipe?

- [ ] A: A plumber
- [ ] B: A pipefitter
- [ ] C: A skilled welder
- [ ] D: A competent, CORGI-registered engineer

22.29

When using a blowtorch to joint copper tube and fittings in a property, a fire extinguisher should be:

- [ ] A: available in the immediate work area
- [ ] B: held over the joint while you are using the blowtorch
- [ ] C: used to cool the fitting
- [ ] D: available only if a property is occupied

22.30

When using a blowtorch, you should stop using the blowtorch:

- [ ] A: immediately before leaving the job:
- [ ] B: at least 1 hour before leaving the job
- [ ] C: at least 2 hours before leaving the job
- [ ] D: at least 4 hours before leaving the job

22.31

When using a blowtorch near to hair-felt lagging, you should:

- [ ] A: just remove enough lagging to carry out the job
- [ ] B: remove the lagging at least 1 metre either side of the work
- [ ] C: remove the lagging at least 3 metres either side of the work
- [ ] D: wet the lagging but leave it in place

22.32

When using a blowtorch near to timber, you should:

- [ ] A: carry out the work taking care not to set fire to the timber
- [ ] B: use a non-combustible mat and have a fire extinguisher ready
- [ ] C: wet the timber first and have a bucket of water handy
- [ ] D: point the flame away from the timber and have a bucket of sand ready to put out the fire

22.33

You are required to take up a length of floorboard to install pipework. Which of the following tools should you use?

- [ ] A: A hammer and wood chisel
- [ ] B: A hammer and screwdriver
- [ ] C: A hammer and bolster
- [ ] D: A chainsaw

Answers: 22.28 = D, 22.29 = A, 22.30 = B, 22.31 = B, 22.32 = B, 22.33 = C

22.34

When using pipe-freezing equipment to isolate the damaged section of pipe, you should:

- [ ] A: always work in pairs when using pipe-freezing equipment
- [ ] B: never allow the freezing gas to come into direct contact with surface water
- [ ] C: never use pipe-freezing equipment on plastic pipe
- [ ] D: wear gloves to avoid direct contact with the skin and read the COSHH assessment

22.35

When drilling a hole for a boiler flue, from the outside of a property at first floor level, which of the following means of access should you use?

- [ ] A: A long ladder
- [ ] B: Borrow some scaffolding and erect it
- [ ] C: A mobile tower
- [ ] D: Packing cases to stand on

22.36

You have to drill through a wall panel that you suspect contains an asbestos material. What should you do?

- [ ] A: Ignore it and carry on
- [ ] B: Hold your breath and carry on
- [ ] C: Put on a dust mask
- [ ] D: Stop work and report it

22.37

When working on a roof to install a flexible flue liner into an existing chimney, you should:

- [ ] A: work from a roof ladder securely hooked over the ridge
- [ ] B: use a stable working platform, with handrails, around or next to the chimney
- [ ] C: scramble up the roof tiles to get to the chimney
- [ ] D: get your mate to do the job while you hold a rope tied to him

22.38

When carrying a ladder on a vehicle, what is the correct way of securing the ladder to the roof rack?

- [ ] A: Rope
- [ ] B: Bungee elastics
- [ ] C: Ladder clamps
- [ ] D: Any of the other answers

Answers: 22.34 = D, 22.35 = C, 22.36 = D, 22.37 = B, 22.38 = C

22.39

The legionella bacteria that cause Legionnaire's disease are most likely to be found in which of the following?

- [ ] A: A boiler operating at a temperature of 80° centigrade
- [ ] B: An infrequently used shower hose outlet
- [ ] C: A cold water storage cistern containing water at 10° centigrade
- [ ] D: A toilet pan

22.40

How are legionella bacteria passed on to humans?

- [ ] A: Through fine water droplets such as sprays or mists
- [ ] B: By drinking dirty water
- [ ] C: Through contact with the skin
- [ ] D: From other people when they sneeze

22.41

Temporary continuity bonding is carried out before removing and replacing sections of metallic pipework, to:

- [ ] A: provide a continuous earth for the pipework installation
- [ ] B: prevent any chance of blowing a fuse
- [ ] C: maintain the live supply to the electrical circuit
- [ ] D: prevent any chance of corrosion to the pipework

22.42

Which type of power drill is most suitable for fixing a run of pipework outside in wet weather?

- [ ] A: Battery-powered drill
- [ ] B: Drill with 110 volt power supply
- [ ] C: Drill with 240 volt power supply
- [ ] D: Any mains voltage drill with a power breaker

22.43

What would you use to find out whether a wall into which you are about to drill contains an electric supply?

- [ ] A: A neon screwdriver
- [ ] B: A cable tracer
- [ ] C: A multimeter
- [ ] D: A hammer and chisel

Answers: 22.39 = B, 22.40 = A, 22.41 = A, 22.42 = A, 22.43 = B

22.44

What is the colour of propane gas cylinders?

- [ ] A: Black
- [ ] B: Maroon
- [ ] C: Red / Orange
- [ ] D: Blue

22.45

Which is the safest method of taking long lengths of copper pipe by van?

- [ ] A: Tying the pipes to the roof with copper wire
- [ ] B: Someone holding the pipes on the roof rack as you drive along
- [ ] C: Putting the pipes inside the van with the ends out of the passenger window
- [ ] D: Using a pipe rack fixed to the roof of the van

22.46

Stepladders must only be used:

- [ ] A: inside buildings
- [ ] B: if they are in good condition and suitable
- [ ] C: if they are made of aluminium
- [ ] D: if they are less than 1.75 metres high

22.47

You are asked to move a cast-iron boiler some distance. What is the safest method?

- [ ] A: Get a workmate to carry it with you
- [ ] B: Drag it
- [ ] C: Roll it end-over-end
- [ ] D: Use a trolley or other manual handling aid

22.48

When drilling a hole through a wall you need to wear eye protection:

- [ ] A: when drilling overhead only
- [ ] B: when the drill bit exceeds 20 mm
- [ ] C: always, whatever the circumstances
- [ ] D: when drilling through concrete only

22.49

What Personal Protective Equipment should you wear when using a hammer drill to drill a 100mm diameter hole through a brick wall?

- [ ] A: Gloves, breathing apparatus and boots
- [ ] B: Ear defenders, face mask and boots
- [ ] C: Ear defenders, breathing apparatus and barrier cream
- [ ] D: Barrier cream, boots and face mask

Answers: 22.44 = C, 22.45 = D, 22.46 = B, 22.47 = D, 22.48 = C, 22.49 = B

22.50

During a job you may need to work below a ground-level suspended timber floor. What is the first question you should ask?

- [ ] A: Can the work be performed from outside?
- [ ] B: Will temporary lighting be used?
- [ ] C: What is contained under the floor?
- [ ] D: How many ways in or out are there?

Answer: 22.50 = A

HVACR: Pipefitting / welding (industrial and commercial)

23.1

If you find a coloured wire sticking out of an electrical plug, which is the correct action to take?

- [ ] A: Push it back into the plug and carry on working
- [ ] B: Pull the wire clear of the plug and report it to your supervisor
- [ ] C: Mark the item as defective and follow your company procedure for defective items
- [ ] D: Take the plug apart and carry out a repair

23.2

Extension leads in use on a site should be:

- [ ] A: located so as to prevent a tripping hazard
- [ ] B: laid out in the shortest, most convenient route
- [ ] C: coiled on a drum or cable tidy
- [ ] D: raised on bricks

23.3

What should you do if you need additional temporary wiring for your power tools whilst working on site?

- [ ] A: Find some cable and extend the wiring yourself
- [ ] B: Stop work until an authorised supply has been installed
- [ ] C: Speak to an electrician and ask him to do the temporary wiring
- [ ] D: Disconnect a longer cable serving somewhere else and reconnect it to where you need it

23.4

When planning a lifting operation, the sequence of operations to enable a lift to be carried out safely should be confirmed in:

- [ ] A: verbal instructions
- [ ] B: a method statement
- [ ] C: a radio telephone message
- [ ] D: a notice in the canteen

Answers: 23.1 = C, 23.2 = A, 23.3 = B, 23.4 = B

23.5

The safe working load (SWL) of lifting equipment is:

- [ ] A: never marked on the equipment but kept with the test certificates
- [ ] B: provided for guidance only
- [ ] C: may be exceeded by no more than 25%
- [ ] D: the absolute maximum safe working load

23.6

Before using a ladder you must make sure that:

- [ ] A: it is secured to prevent it from moving sideways or sliding outwards
- [ ] B: no one else has booked the ladder for their work
- [ ] C: an apprentice or workmate is standing by in case you slip and fall
- [ ] D: the weather forecast is for a bright, clear day

23.7

When positioning and erecting a stepladder, which of the following is essential for its safe use?

- [ ] A: It has a tool tray towards the top of the steps
- [ ] B: The restraint mechanism is spread to its full extent
- [ ] C: You will be able to reach the job by standing on the top step
- [ ] D: A competent person has positioned and erected the steps

23.8

Generally, how many working platforms should be in use at any one time on a mobile tower?

- [ ] A: One
- [ ] B: Two
- [ ] C: Three
- [ ] D: Four

23.9

What is the recommended maximum height for a free-standing mobile tower?

- [ ] A: There is no restriction
- [ ] B: 2 metres
- [ ] C: In accordance with the manufacturer's recommendations
- [ ] D: 12 metres

Answers: 23.5 = D, 23.6 = A, 23.7 = B, 23.8 = A, 23.9 = C

23.10

What is the first thing you should do after getting on to the platform of a mobile tower?

- [ ] A: Check that the brakes are locked on
- [ ] B: Check the mobile tower to make sure that it has been correctly assembled
- [ ] C: Close the access hatch to prevent falls of personnel, tools or equipment
- [ ] D: Make sure that the tower does not rock or wobble

23.11

What should be done before a mobile tower is moved?

- [ ] A: All people and equipment must be removed from the platform
- [ ] B: A Permit to Work is required
- [ ] C: The Principal Contractor must give their approval
- [ ] D: Arrangements must be made with the forklift truck driver

23.12

What must be done first before any roof work is carried out?

- [ ] A: A risk assessment must be carried out
- [ ] B: The operatives working on the roof must be trained in the use of safety harnesses
- [ ] C: Permits to Work must be issued only to those allowed to work on the roof
- [ ] D: A weather forecast must be obtained

23.13

What is edge protection designed to do?

- [ ] A: Make access to the roof easier
- [ ] B: Secure tools and materials close to the edge
- [ ] C: Prevent rainwater running off the roof onto workers below
- [ ] D: Prevent the fall of people and materials

Answers: 23.10 = C, 23.11 = A, 23.12 = A, 23.13 = D

23.14

Where should Liquefied Petroleum Gas (LPG) cylinders be positioned when supplying an appliance in a site cabin?

- [ ] A: Inside the site cabin in a locked cupboard
- [ ] B: Under the cabin
- [ ] C: Inside the cabin next to the appliance
- [ ] D: Outside the cabin

23.15

If you are working where welding is being carried out, what should be provided to protect you from 'welding flash'?

- [ ] A: Fire extinguishers
- [ ] B: Warning notices
- [ ] C: Screens
- [ ] D: High visibility vest

23.16

How should you position the exhaust of an engine-driven generator which has to be run inside a building?

- [ ] A: Position the exhaust outside the building
- [ ] B: Position the exhaust in a stairwell
- [ ] C: Hang the exhaust in another room
- [ ] D: Position the exhaust in a riser

23.17

How should cylinders containing Liquefied Petroleum Gas (LPG) be stored on site?

- [ ] A: In a locked cellar with clear warning signs
- [ ] B: In a locked external compound at least 3 metres from any oxygen cylinders
- [ ] C: As close to the point of use as possible
- [ ] D: Covered by a tarpaulin to shield the compressed cylinder from sunlight

23.18

When working in a riser, how should access be controlled?

- [ ] A: By a site security operative
- [ ] B: By those who are working in it
- [ ] C: By the main contractor
- [ ] D: By a Permit to Work system

23.19

Which of the following electrical equipment does not require portable appliance testing?

- [ ] A: Battery-powered rechargeable drill
- [ ] B: 110 volt electrical drill
- [ ] C: 110 volt portable halogen light
- [ ] D: Electric kettle

Answers: 23.14 = D, 23.15 = C, 23.16 = A, 23.17 = B, 23.18 = D, 23.19 = A

23.20

What must be clearly marked on all lifting equipment?

- [ ] A: Name of the manufacturer
- [ ] B: The safe working load
- [ ] C: Next test date
- [ ] D: Specification of material from which made

23.21

What is the first thing you should do after getting on to the platform of a correctly erected mobile tower?

- [ ] A: Check that the brakes are locked on
- [ ] B: Check the mobile tower to make sure that it has been correctly assembled
- [ ] C: Close the access hatch to prevent falls of personnel, tools or equipment
- [ ] D: Make sure that the tower does not rock or wobble

23.22

When assembling a mobile tower what major overhead hazard must you be aware of?

- [ ] A: Water pipes
- [ ] B: Cable trays
- [ ] C: False ceilings
- [ ] D: Suspended electric cables

23.23

You spill some oil on the floor and you do not have any absorbent material to clean the area. What should you do?

- [ ] A: Spread it about to lessen the depth
- [ ] B: Keep people out of the area and inform your supervisor
- [ ] C: Do nothing, it will eventually soak into the floor
- [ ] D: Warn other people as they tread through it

23.24

What should folding stepladders be used for?

- [ ] A: General access on site
- [ ] B: Short-term activities lasting minutes that don't involve stretching
- [ ] C: All site activities where a straight ladder cannot be used
- [ ] D: Getting on and off mobile towers

Answers: 23.20 = B, 23.21 = C, 23.22 = D, 23.23 = B, 23.24 = B

23.25

When a new piece of plant has been installed but has not been commissioned, how should it be left?

- [ ] A: With any valves and switches turned off
- [ ] B: With any valves and switches clearly labelled
- [ ] C: With all valves and switches 'locked off'
- [ ] D: With any valves and switches turned on and ready to use

23.26

What is the main hazard associated with flame-cutting and welding?

- [ ] A: Gas poisoning
- [ ] B: Fire
- [ ] C: Dropping a gas cylinder
- [ ] D: Not having a 'Hot Work' permit

23.27

What guarding is required when a pipe threading machine is in use?

- [ ] A: A length of red material hung from the exposed end of the pipe
- [ ] B: A barrier at the exposed end of the pipe only
- [ ] C: A barrier around the whole length of the pipe
- [ ] D: Warning notices in the work area

23.28

When using a blowtorch, you should:

- [ ] A: stop using the blowtorch immediately before leaving the job
- [ ] B: stop using the blowtorch at least 1 hour before leaving the job
- [ ] C: stop using the blowtorch at least 2 hours before leaving the job
- [ ] D: stop using the blowtorch at least 4 hours before leaving the job

23.29

When using a blowtorch near to hair-felt lagging you should:

- [ ] A: just remove enough lagging to carry out the work
- [ ] B: remove the lagging at least 1 metre either side of the work
- [ ] C: remove the lagging at least 3 metres either side of the work
- [ ] D: wet the lagging but leave it in place

Answers: 23.25 = C, 23.26 = B, 23.27 = C, 23.28 = B, 23.29 = B

23.30

When using a blowtorch near to timber, you should:

- [ ] A: carry out the work taking care not to catch the timber
- [ ] B: use a non-combustible mat and have a fire extinguisher ready
- [ ] C: wet the timber first and keep a bucket of water handy
- [ ] D: point the flame away from the timber and have a bucket of sand ready to put out the fire

23.31

When using pipe-freezing equipment to isolate the damaged section of pipe, you should:

- [ ] A: always work in pairs
- [ ] B: never allow the freezing gas to come into direct contact with surface water
- [ ] C: never use pipe-freezing equipment on plastic pipe
- [ ] D: wear gloves to avoid direct contact with the skin and read the COSHH assessment

23.32

What is the colour of an acetylene cylinder?

- [ ] A: Orange
- [ ] B: Black
- [ ] C: Green
- [ ] D: Maroon

23.33

What particular item of personal protective equipment should you use when oxyacetylene brazing?

- [ ] A: Ear defenders
- [ ] B: Clear goggles
- [ ] C: Green-tinted goggles
- [ ] D: Dust mask

23.34

When using oxyacetylene brazing equipment the bottles should be:

- [ ] A: laid on their side and secured
- [ ] B: stood upright and secured
- [ ] C: stood upside down
- [ ] D: angled at 45° and secured against falling

Answers: 23.30 = B, 23.31 = D, 23.32 = D, 23.33 = C, 23.34 = B

23.35

You are asked to install high-level pipework from a platform that has no edge protection and is located above an open stairwell. Do you:

- [ ] A: get on with the job, but keep away from the edge of the platform
- [ ] B: not start work until your work platform has been fitted with guard-rails and toe-boards
- [ ] C: get on with the job, ensuring that a workmate stays close by
- [ ] D: get on with the job, provided that if you fall the stairwell guard-rail will prevent you from falling further

23.36

Which two of the following are essential safety checks to be carried out before using oxyacetylene equipment?

- [ ] A: The cylinders are full
- [ ] B: The cylinders, hoses and flashback arresters are in good condition
- [ ] C: The trolley wheels are the right size
- [ ] D: The area is well ventilated and clear of any obstructions
- [ ] E: The cylinders are the right weight

23.37

Who is allowed to install natural gas pipework?

- [ ] A: A skilled engineer
- [ ] B: A pipefitter
- [ ] C: A CORGI-registered engineer
- [ ] D: Anybody

23.38

Who should carry out pressure testing on pipework or vessels?

- [ ] A: Anyone who is available
- [ ] B: A competent person
- [ ] C: A Health and Safety Executive inspector
- [ ] D: A Building Control Officer

23.39

During the pressure testing of pipework or vessels, who should be present?

- [ ] A: The architect
- [ ] B: The site foreman
- [ ] C: Only those involved in carrying out the test
- [ ] D: Anybody

Answers: 23.35 = B, 23.36 = B,D, 23.37 = C, 23.38 = B, 23.39 = C

23.40

When using an electrically powered threading machine you should make sure that:

- [ ] A: the power supply is 24 volts
- [ ] B: the power supply is 415 volts and the machine fitted with a guard
- [ ] C: your clothing cannot get caught on rotating parts of the machine
- [ ] D: you wear ear defenders

23.41

Why is it important to know the difference between propane and butane equipment?

- [ ] A: Propane equipment operates at higher pressure
- [ ] B: Propane equipment operates at lower pressure
- [ ] C: Propane equipment is cheaper
- [ ] D: Propane equipment can be used with smaller, easy-to-handle cylinders

23.42

Which of the following statements is true?

- [ ] A: Both propane and butane are heavier than air
- [ ] B: Butane is heavier than air while propane is lighter than air
- [ ] C: Propane is heavier than air while butane is lighter than air
- [ ] D: Both propane and butane are lighter than air

23.43

Apart from the cylinders used in gas-powered forklift trucks, Liquefied Petroleum Gas (LPG) cylinders should never be placed on their sides during use because:

- [ ] A: it would give a faulty reading on the contents gauge, resulting in flashback
- [ ] B: air could be drawn into the cylinder, creating a dangerous mixture of gases
- [ ] C: the liquid gas would be at too low a level to allow the torch to burn correctly
- [ ] D: the liquid gas could be drawn from the cylinder, creating a safety hazard

Answers: 23.40 = C, 23.41 = A, 23.42 = A, 23.43 = D

23.44

What is the method of checking for leaks after connecting a Liquefied Petroleum Gas (LPG) regulator to the bottle?

- [ ] A: Test with a lighted match
- [ ] B: Sniff the connections to detect the smell of gas
- [ ] C: Listen to hear for escaping gas
- [ ] D: Apply leak detection fluid to the connections

23.45

What is the most likely risk of injury when cutting a pipe with hand-operated pipe cutters?

- [ ] A: Your fingers may become trapped between the cutting wheel and the pipe
- [ ] B: The inside edge of the cut pipe becomes extremely sharp to touch
- [ ] C: Continued use can cause muscle damage
- [ ] D: Pieces of sharp metal could fly off

23.46

The use of oxyacetylene equipment is not recommended for which of the following jointing methods?

- [ ] A: Jointing copper pipe using hard soldering
- [ ] B: Jointing copper tube using capillary soldered fittings
- [ ] C: Jointing mild steel tube
- [ ] D: Jointing sheet lead

23.47

Why is it essential to take great care when handling oxygen cylinders?

- [ ] A: They contain highly flammable compressed gas
- [ ] B: They contain highly flammable liquid gas
- [ ] C: They are filled to extremely high pressures
- [ ] D: They contain poisonous gas

23.48

Where should acetylene gas welding bottles be stored when they are not in use?

- [ ] A: Outside in a special storage compound
- [ ] B: In special rack in a company van
- [ ] C: Inside a building in a locked cupboard
- [ ] D: With oxygen bottles

Answers: 23.44 = D, 22.45 = B, 22.46 = B, 22.47 = C, 22.48 = A

23.49

When drilling a hole through a wall you need to wear eye protection when:

- [ ] A: drilling overhead only
- [ ] B: the drill bit exceeds 20 mm
- [ ] C: always, whatever the circumstances
- [ ] D: drilling through concrete only

23.50

While working on your own and tracing pipework in a building, the pipes enter a service duct. You should:

- [ ] A: go into the service duct and continue to trace the pipework
- [ ] B: ask someone in the building to act as your second person
- [ ] C: put on your Personal Protective Equipment and carry on with the job
- [ ] D: stop work until a risk assessment has been carried out

Answers: 23.49 = C, 23.50 = D

24.1

If you find a coloured wire sticking out of an electrical plug, which is the correct action to take?

- [ ] A: Push it back into the plug and carry on working
- [ ] B: Pull the wire clear of the plug and report it to your supervisor
- [ ] C: Mark the item as defective and follow your company procedure for defective items
- [ ] D: Take the plug apart and carry out a repair

24.2

Extension leads in use on a site should be:

- [ ] A: located so as to prevent a tripping hazard
- [ ] B: laid out in the shortest, most convenient route
- [ ] C: coiled on a drum or cable tidy
- [ ] D: raised on bricks

24.3

What should you do if you need additional temporary wiring for your power tools whilst working on site?

- [ ] A: Find some cable and extend the wiring yourself
- [ ] B: Stop work until an authorised supply has been installed
- [ ] C: Speak to an electrician and ask him to do the temporary wiring
- [ ] D: Disconnect a longer cable serving somewhere else and reconnect it to where you need it

24.4

When planning a lifting operation the sequence of operations to enable a lift to be carried out safely should be confirmed in:

- [ ] A: verbal instructions
- [ ] B: a method statement
- [ ] C: a radio telephone message
- [ ] D: a notice in the canteen

Answers: 24.1 = C, 24.2 = A, 24.3 = B, 24.4 = B

24.5

The safe working load (SWL) of lifting equipment is:

- [ ] A: never marked on the equipment but kept with the test certificates
- [ ] B: provided for guidance only
- [ ] C: may be exceeded by no more than 25%
- [ ] D: the absolute maximum safe working load

24.6

Before using a ladder you must make sure that:

- [ ] A: it is secured to prevent it from moving sideways or sliding outwards
- [ ] B: no one else has booked the ladder for their work
- [ ] C: an apprentice or workmate is standing by in case you slip and fall
- [ ] D: the weather forecast is for a bright, clear day

24.7

When positioning and erecting a stepladder, which of the following is essential for its safe use?

- [ ] A: It has a tool tray towards the top of the steps
- [ ] B: The restraint mechanism is spread to its full extent
- [ ] C: You will be able to reach the job by standing on the top step
- [ ] D: A competent person has positioned and erected the steps

24.8

Generally, how many working platforms should be in use at any one time on a mobile tower?

- [ ] A: One
- [ ] B: Two
- [ ] C: Three
- [ ] D: Four

24.9

What is the recommended maximum height for a free-standing mobile tower?

- [ ] A: There is no restriction
- [ ] B: 2 metres
- [ ] C: In accordance with the manufacturer's recommendations
- [ ] D: 12 metres

Answers: 24.5 = D, 24.6 = A, 24.7 = B, 24.8 = A, 24.9 = C

24.10

What is the first thing you should do after getting on to the platform of a mobile tower?

- [ ] A: Check that the brakes are locked on
- [ ] B: Check the mobile tower to make sure that it has been correctly assembled
- [ ] C: Close the access hatch to prevent falls of personnel, tools or equipment
- [ ] D: Make sure that the tower does not rock or wobble

24.11

What should be done before a mobile tower is moved?

- [ ] A: All people and equipment must be removed from the platform
- [ ] B: A Permit to Work is required
- [ ] C: The Principal Contractor must give their approval
- [ ] D: Arrangements must be made with the forklift truck driver

24.12

What must be done first before any roof work is carried out?

- [ ] A: A risk assessment must be carried out
- [ ] B: The operatives working on the roof must be trained in the use of safety harnesses
- [ ] C: Permits to Work must be issued only to those allowed to work on the roof
- [ ] D: A weather forecast must be obtained

24.13

What is edge protection designed to do?

- [ ] A: Make access to the roof easier
- [ ] B: Secure tools and materials close to the edge
- [ ] C: Prevent rainwater running off the roof onto workers below
- [ ] D: Prevent the fall of people and materials

Answers: 24.10 = C, 24.11 = A, 24.12 = A, 24.13 = D

24.14

Where should Liquefied Petroleum Gas (LPG) cylinders be positioned when supplying an appliance in a site cabin?

- [ ] A: Inside the site cabin in a locked cupboard
- [ ] B: Under the cabin
- [ ] C: Inside the cabin next to the appliance
- [ ] D: Outside the cabin

24.15

If you are working where welding is being carried out, what should be provided to protect you from 'welding flash'?

- [ ] A: Fire extinguishers
- [ ] B: Warning notices
- [ ] C: Screens
- [ ] D: High visibility vest

24.16

How should you position the exhaust of an engine-driven generator which has to be run inside a building?

- [ ] A: Position the exhaust outside the building
- [ ] B: Position the exhaust in a stairwell
- [ ] C: Hang the exhaust in another room
- [ ] D: Position the exhaust in a riser

24.17

How should cylinders containing Liquefied Petroleum Gas (LPG) be stored on site?

- [ ] A: In a locked cellar with clear warning signs
- [ ] B: In a locked external compound at least 3 metres from any oxygen cylinders
- [ ] C: As close to the point of use as possible
- [ ] D: Covered by a tarpaulin to shield the compressed cylinder from sunlight

24.18

When working in a riser, how should access be controlled?

- [ ] A: By a site security operative
- [ ] B: By those who are working in it
- [ ] C: By the main contractor
- [ ] D: By a Permit to Work system

24.19

Which of the following electrical equipment does not require portable appliance testing?

- [ ] A: Battery-powered rechargeable drill
- [ ] B: 110 volt electrical drill
- [ ] C: 110 volt portable halogen light
- [ ] D: Electric kettle

Answers: 24.14 = D, 24.15 = C, 24.16 = A, 24.17 = B, 24.18 = D, 24.19 = A

24.20

What must be clearly marked on all lifting equipment?

- [ ] A: Name of the manufacturer
- [ ] B: The safe working load
- [ ] C: Next test date
- [ ] D: Specification of material from which made

24.21

What is the first thing you should do after getting on to the platform of a correctly erected mobile tower?

- [ ] A: Check that the brakes are locked on
- [ ] B: Check the mobile tower to make sure that it has been correctly assembled
- [ ] C: Close the access hatch to prevent falls of personnel, tools or equipment
- [ ] D: Make sure that the tower does not rock or wobble

24.22

When assembling a mobile tower what major overhead hazard must you be aware of?

- [ ] A: Water pipes
- [ ] B: Cable trays
- [ ] C: False ceilings
- [ ] D: Suspended electric cables

24.23

You spill some oil on the floor and you do not have any absorbent material to clean the area. What should you do?

- [ ] A: Spread it about to lessen the depth
- [ ] B: Keep people out of the area and inform your supervisor
- [ ] C: Do nothing, it will eventually soak into the floor
- [ ] D: Warn other people as they tread through it

24.24

What should folding stepladders be used for?

- [ ] A: General access on site
- [ ] B: Short-term activities lasting minutes that don't involve stretching
- [ ] C: All site activities where a straight ladder cannot be used
- [ ] D: Getting on and off mobile towers

Answers: 24.20 = B, 24.21 = C, 24.22 = D, 24.23 = B, 24.24 = B

24.25

When a new piece of plant has been installed but has not been commissioned, how should it be left?

- [ ] A: With any valves and switches turned off
- [ ] B: With any valves and switches clearly labelled
- [ ] C: With all valves and switches 'locked off'
- [ ] D: With any valves and switches turned on and ready to use

24.26

How should you leave the ends of ductwork after using a solvent-based sealant?

- [ ] A: Seal up all the open ends to ensure that dirt cannot get into the system
- [ ] B: Ensure that the lids are left off tins of solvent
- [ ] C: Remove any safety signs or notices
- [ ] D: Leave inspection covers off and erect 'No smoking' signs

24.27

You are working on a refurbishment and removing some old ductwork when you notice that it has been insulated using a white, powdery material that could be asbestos. What should you do?

- [ ] A: Rip off the insulation as quickly as possible and place it in the skip
- [ ] B: Get a face mask for yourself and get some assistance to do the job quickly
- [ ] C: Stop work immediately and report it
- [ ] D: Take off the material carefully and place it in a sealed container

24.28

When using a material hoist you notice that the lifting cable is frayed. What should you do?

- [ ] A: Get the job done as fast as possible
- [ ] B: Straighten out the cable using mole grips
- [ ] C: Do not use the hoist and report the problem
- [ ] D: Be very careful when using the hoist

Answers: 24.25 = C, 24.26 = D, 24.27= C, 24.28 = C

24.29

Before taking down a run of ductwork, what is the first thing you should do?

- [ ] A: Get a big skip to put it in
- [ ] B: Cut through the support rods
- [ ] C: Clean the ductwork to remove all dust
- [ ] D: Assess the task to be undertaken

24.30

You are asked to move a fan-coil unit some distance. What is the safest way to do it?

- [ ] A: Get a workmate to carry it with you
- [ ] B: Drag it
- [ ] C: Roll it end-over-end
- [ ] D: Use a trolley or other manual handling aid

24.31

When drilling a hole through a wall you need to wear eye protection:

- [ ] A: when drilling overhead only
- [ ] B: when the drill bit exceeds 20 mm
- [ ] C: always, whatever the circumstances
- [ ] D: when drilling through concrete only

24.32

In addition to a safety helmet and protective footwear, what Personal Protective Equipment should you wear when using a hammer drill?

- [ ] A: Gloves and breathing apparatus
- [ ] B: Hearing protection, face mask and eye protection
- [ ] C: Hearing protection, breathing apparatus and barrier cream
- [ ] D: Barrier cream and face mask

24.33

What should you do if you see the side of an abrasive disc on a disc-cutter being used to grind down the ends of a drop rod?

- [ ] A: Check that the correct type of disc is being used
- [ ] B: Stop the person immediately
- [ ] C: Stop work and report it
- [ ] D: Check that the disc has not been subjected to over-speeding

Answers: 24.29 = D, 24.30 = D, 24.31 = C, 24.32 = B, 24.33 = B

24.34

While fitting a fire damper into a ductwork system you notice that, due to a manufacturing fault, it may not operate properly. You should:

- [ ] A: install it anyway, as it is
- [ ] B: fix it so that it stays open, and then install it
- [ ] C: not fit the damper and report the fault
- [ ] D: leave it out of the ductwork system altogether

24.35

You are asked to install high-level ductwork from a platform that has no edge protection and is located above an open stairwell. You should:

- [ ] A: get on with the job, but keep away from the edge of the platform
- [ ] B: not start work until your work platform has been fitted with guard-rails and toe-boards
- [ ] C: get on with the job, ensuring that a workmate stays close by
- [ ] D: get on with the job, provided that if you fall the stairwell guard-rail will prevent you from falling further

24.36

You have to cut some flexible aluminium ductwork that is pre-insulated with a fibreglass material. Which two of the following should you use?

- [ ] A: A hacksaw
- [ ] B: Respiratory protection
- [ ] C: A disc cutter
- [ ] D: A set of tin snips
- [ ] E: Ear defenders

24.37

You have been asked to install a run of ceiling-mounted ductwork across a large open-plan area, which has a good floor. It is best to do this:

- [ ] A: using stepladders
- [ ] B: using scaffold boards and floor stands
- [ ] C: using packing cases to stand on
- [ ] D: from mobile access towers fitted with guard-rails and toe-boards

Answers: 24.34 = C, 24.35 = B, 24.36 = B,D, 24.37 = D

24.38

You have to dismantle some waste-extract ductwork. What is the first thing you should do?

- [ ] A: Arrange for a skip to put it in
- [ ] B: Find out what the ductwork may be contaminated with
- [ ] C: Check that the duct supports are strong enough to cope with the dismantling
- [ ] D: Make sure there are enough disc cutters to do the job

24.39

You have to carry out a job on the flat roof of a two-storey building, about one metre from the edge of the roof, which has a very low parapet. You should:

- [ ] A: carry on with the job, provided that you don't get dizzy with heights
- [ ] B: use a full body harness, lanyard and anchor while doing the job
- [ ] C: ask for double guard-rails and a toe-board to be installed to prevent you falling
- [ ] D: get your mate to do the work, while you hold on to him

24.40

While using a 'genie' hoist you notice that part of the hoist is buckling slightly. You should:

- [ ] A: lower the load immediately
- [ ] B: carry on with the job, while keeping an eye on the buckling metal
- [ ] C: straighten out the buckled metal and then get on with the lifting operation
- [ ] D: get the job finished quickly

24.41

A person who has been using a solvent-based ductwork sealant is complaining of headaches and feeling sick. What is the first thing you should do?

- [ ] A: Let them carry on working but try to keep a close watch on them
- [ ] B: Get them a drink of water and a headache tablet
- [ ] C: Get them out to fresh air and make them rest
- [ ] D: Stop work and tell your supervisor

Answers: 24.38 = B, 24.48 = C, 24.40 = A, 24.41 = C

24.42

What additional control measure must be put in place when welding in-situ galvanised ductwork?

- [ ] A: Screens
- [ ] B: Fume extraction
- [ ] C: Warning signs
- [ ] D: Hearing protection

24.43

When carrying out solvent welding on plastic ductwork, what particular safety measure must be applied?

- [ ] A: The area must be well ventilated
- [ ] B: The supervisor must be present
- [ ] C: A hard hat must be worn
- [ ] D: It must be done in daylight

24.44

Which of the following need you not do before using a cleaning agent or biocide in a ductwork system?

- [ ] A: Ask for advice from the cleaning agent or biocide manufacturer
- [ ] B: Read the COSHH assessment for the material, carrying out a risk assessment and producing a method statement for the work
- [ ] C: Consult the building occupier
- [ ] D: Check on what the ductwork will carry in the future

24.45

Which of the following need not be done before cleaning a system in industrial, laboratory or other premises where you might encounter harmful particulates?

- [ ] A: Examine the system
- [ ] B: Collect a sample from the ductwork
- [ ] C: Run the system under overload conditions
- [ ] D: Prepare a job-specific risk assessment and method statement

24.46

Where it is necessary to enter ductwork, which are the two main factors that need to be considered?

- [ ] A: Working in a confined space
- [ ] B: What the ductwork will carry in the future
- [ ] C: The cleanliness of the ductwork
- [ ] D: Wearing kneepads
- [ ] E: The strength of the ductwork and its supports

Answers: 24.42 = B, 24.43 = A, 24.44 = D, 24.45 = C, 24.46 = A,E

24.47

What particular factor should be considered before working on a kitchen extraction system?

- [ ] A: Access to the ductwork
- [ ] B: The cooking deposits within the ductwork
- [ ] C: The effect on the future performance of the system
- [ ] D: The effect on food preparation

24.48

What should you do before painting the external surface of ductwork?

- [ ] A: Clean the paintbrushes
- [ ] B: Read the COSHH assessment before using the paint
- [ ] C: Switch off the system
- [ ] D: Put on eye protection

24.49

When jointing plastic-coated metal ductwork, which of the following methods of jointing presents the most serious risk to health?

- [ ] A: Welding
- [ ] B: Taping
- [ ] C: Riveting
- [ ] D: Nuts and bolts

24.50

Who should carry out leakage testing of a newly installed ductwork system?

- [ ] A: The installation contractor
- [ ] B: The property owner
- [ ] C: The designer
- [ ] D: A trained and competent person

24.51

You are removing a run of ductwork in an unoccupied building and notice a hypodermic syringe behind it. What should you do?

- [ ] A: Ensure the syringe is empty, remove the syringe and place it with the rubbish
- [ ] B: Wear gloves, break the syringe into small pieces and flush it down the drain
- [ ] C: Notify the supervisor, cordon off the area and call the emergency services
- [ ] D: Wearing gloves, use grips to remove the syringe to a safe place and report your find

Answers: 24.47 = B, 24.48 = B, 24.49 = A, 24.50 = D, 24.51 = D

HVACR: Refrigeration and Air Conditioning

25.1

If you find a coloured wire sticking out of an electrical plug, which is the correct action to take?

- A: Push it back into the plug and carry on working
- B: Pull the wire clear of the plug and report it to your supervisor
- C: Mark the item as defective and follow your company procedure for defective items
- D: Take the plug apart and carry out a repair

25.2

Extension leads in use on a site should be:

- A: located so as to prevent a tripping hazard
- B: laid out in the shortest, most convenient route
- C: coiled on a drum or cable tidy
- D: raised on bricks

25.3

What should you do if you need additional temporary wiring for your power tools whilst working on site?

- A: Find some cable and extend the wiring yourself
- B: Stop work until an authorised supply has been installed
- C: Speak to an electrician and ask him to do the temporary wiring
- D: Disconnect a longer cable serving somewhere else and reconnect it to where you need it

25.4

When planning a lifting operation the sequence of operations to enable a lift to be carried out safely should be confirmed in:

- A: verbal instructions
- B: a method statement
- C: a radio telephone message
- D: a notice in the canteen

Answers: 25.1 = C, 25.2 = A, 25.3 = B, 25.4 = B

25.5

The safe working load (SWL) of lifting equipment is:

- A: never marked on the equipment but kept with the test certificates
- B: provided for guidance only
- C: may be exceeded by no more than 25%
- D: the absolute maximum safe working load

25.6

Before using a ladder you must make sure that:

- A: it is secured to prevent it from moving sideways or sliding outwards
- B: no one else has booked the ladder for their work
- C: an apprentice or workmate is standing by in case you slip and fall
- D: the weather forecast is for a bright, clear day

25.7

When positioning and erecting a stepladder, which of the following is essential for its safe use?

- A: It has a tool tray towards the top of the steps
- B: The restraint mechanism is spread to its full extent
- C: You will be able to reach the job by standing on the top step
- D: A competent person has positioned and erected the steps

25.8

Generally, how many working platforms should be in use at any one time on a mobile tower?

- A: One
- B: Two
- C: Three
- D: Four

25.9

What is the recommended maximum height for a free-standing mobile tower?

- A: There is no restriction
- B: 2 metres
- C: In accordance with the manufacturer's recommendations
- D: 12 metres

Answers: 25.5 = D, 25.6 = A, 25.7 = B, 25.8 = A, 25.9 = C

25.10

What is the first thing you should do after getting on to the platform of a mobile tower?

- [ ] A: Check that the brakes are locked on
- [ ] B: Check the mobile tower to make sure that it has been correctly assembled
- [ ] C: Close the access hatch to prevent falls of personnel, tools or equipment
- [ ] D: Make sure that the tower does not rock or wobble

25.11

What should be done before a mobile tower is moved?

- [ ] A: All people and equipment must be removed from the platform
- [ ] B: A Permit to Work is required
- [ ] C: The Principal Contractor must give their approval
- [ ] D: Arrangements must be made with the forklift truck driver

25.12

What must be done first before any roof work is carried out?

- [ ] A: A risk assessment must be carried out
- [ ] B: The operatives working on the roof must be trained in the use of safety harnesses
- [ ] C: Permits to Work must be issued only to those allowed to work on the roof
- [ ] D: A weather forecast must be obtained

25.13

What is edge protection designed to do?

- [ ] A: Make access to the roof easier
- [ ] B: Secure tools and materials close to the edge
- [ ] C: Prevent rainwater running off the roof onto workers below
- [ ] D: Prevent the fall of people and materials

Answers: 25.10 = C, 25.11 = A, 25.12 = A, 25.13 = D

25.14

Where should Liquefied Petroleum Gas (LPG) cylinders be positioned when supplying an appliance in a site cabin?

- [ ] A: Inside the site cabin in a locked cupboard
- [ ] B: Under the cabin
- [ ] C: Inside the cabin next to the appliance
- [ ] D: Outside the cabin

25.15

If you are working where welding is being carried out, what should be provided to protect you from 'welding flash'?

- [ ] A: Fire extinguishers
- [ ] B: Warning notices
- [ ] C: Screens
- [ ] D: High visibility vest

25.16

How should you position the exhaust of an engine-driven generator which has to be run inside a building?

- [ ] A: Position the exhaust outside the building
- [ ] B: Position the exhaust in a stairwell
- [ ] C: Hang the exhaust in another room
- [ ] D: Position the exhaust in a riser

25.17

How should cylinders containing Liquefied Petroleum Gas (LPG) be stored on site?

- [ ] A: In a locked cellar with clear warning signs
- [ ] B: In a locked external compound at least 3 metres from any oxygen cylinders
- [ ] C: As close to the point of use as possible
- [ ] D: Covered by a tarpaulin to shield the compressed cylinder from sunlight

25.18

When working in a riser, how should access be controlled?

- [ ] A: By a site security operative
- [ ] B: By those who are working in it
- [ ] C: By the main contractor
- [ ] D: By a Permit to Work system

25.19

Which of the following electrical equipment does not require portable appliance testing?

- [ ] A: Battery-powered rechargeable drill
- [ ] B: 110 volt electrical drill
- [ ] C: 110 volt portable halogen light
- [ ] D: Electric kettle

Answers: 25.14 = D, 25.15 = C, 25.16 = A, 25.17 = B, 25.18 = D, 25.19 = A

25.20

What must be clearly marked on all lifting equipment?

- [ ] A: Name of the manufacturer
- [ ] B: The safe working load
- [ ] C: Next test date
- [ ] D: Specification of material from which made

25.21

What is the first thing you should do after getting on to the platform of a correctly erected mobile tower?

- [ ] A: Check that the brakes are locked on
- [ ] B: Check the mobile tower to make sure that it has been correctly assembled
- [ ] C: Close the access hatch to prevent falls of personnel, tools or equipment
- [ ] D: Make sure that the tower does not rock or wobble

25.22

When assembling a mobile tower what major overhead hazard must you be aware of?

- [ ] A: Water pipes
- [ ] B: Cable trays
- [ ] C: False ceilings
- [ ] D: Suspended electric cables

25.23

You spill some oil on the floor and you do not have any absorbent material to clean the area. What should you do?

- [ ] A: Spread it about to lessen the depth
- [ ] B: Keep people out of the area and inform your supervisor
- [ ] C: Do nothing, it will eventually soak into the floor
- [ ] D: Warn other people as they tread through it

25.24

What should folding stepladders be used for?

- [ ] A: General access on site
- [ ] B: Short-term activities lasting minutes that don't involve stretching
- [ ] C: All site activities where a straight ladder cannot be used
- [ ] D: Getting on and off mobile towers

Answers: 25.20 = B, 25.21 = C, 25.22 = D, 25.23 = B, 25.24 = B

25.25

When a new piece of plant has been installed but has not been commissioned, how should it be left?

- [ ] A: With any valves and switches turned off
- [ ] B: With any valves and switches clearly labelled
- [ ] C: With all valves and switches 'locked off'
- [ ] D: With any valves and switches turned on and ready to use

25.26

Who is permitted to work with refrigerant gases?

- [ ] A: A CORGI gas engineer
- [ ] B: The gas board
- [ ] C: A competent, trained person
- [ ] D: An electrician

25.27

What is the colour of an acetylene cylinder?

- [ ] A: Orange
- [ ] B: Black
- [ ] C: Green
- [ ] D: Maroon

25.28

What particular item of personal protective equipment should you use when oxyacetylene brazing?

- [ ] A: Ear defenders
- [ ] B: Clear goggles
- [ ] C: Green-tinted goggles
- [ ] D: Dust mask

25.29

When using oxyacetylene brazing equipment, the bottles should be:

- [ ] A: laid on their side
- [ ] B: stood upright and secured
- [ ] C: stood upside down
- [ ] D: angled at 45°

25.30

Which is the safest place to store refrigerant cylinders when they are not in use?

- [ ] A: Outside in a special locked storage compound
- [ ] B: In a company vehicle
- [ ] C: Inside the building in a locked cupboard
- [ ] D: In the immediate work area, ready for use the next day

Answers: 25.25 = C, 25.26 = C, 25.27 = D, 25.28 = C, 25.29 = B, 25.30 = A

25.31

When working on refrigeration systems containing hydrocarbon (HC) gases, what particular dangers need to be considered?

- [ ] A: There should be no sources of ignition
- [ ] B: Special Personal Protective Equipment should be worn
- [ ] C: Extra lighting is needed
- [ ] D: The work cannot be carried out when the weather is hot

25.32

Which TWO of the following are essential safety checks to be carried out before oxyacetylene equipment is used?

- [ ] A: The cylinders are full
- [ ] B: The cylinders, hoses and flashback arresters are in good condition
- [ ] C: The trolley wheels are the right size
- [ ] D: The area is well ventilated and clear of any obstructions
- [ ] E: The cylinders are the right weight

25.33

If refrigerant gases are released into a closed room in a building, they would:

- [ ] A: sink to the floor
- [ ] B: rise to the celing
- [ ] C: stay at the same level
- [ ] D: disperse safely within the room

25.34

When handling refrigerant gases, what Personal Protective Equipment should you wear?

- [ ] A: Eye protection, gloves, helmet, overalls
- [ ] B: Gloves, overalls, safety boots, eye protection
- [ ] C: Safety boots, eye protection, harness, overalls
- [ ] D: Overalls, gloves, helmet, safety boots

Answers: 25.31 = A, 25.32 = B,D, 25.33 = A, 35.34 = B

25.35

You have to carry out an installation on the flat roof of a single-storey building, about one metre from the edge of the roof, which has no parapet. You should:

- [ ] A: carry on with the job, provided you don't get dizzy with heights
- [ ] B: use a full body harness, lanyard and anchor while doing the job
- [ ] C: ask for double guard-rails and toe-boards to be installed to prevent you falling
- [ ] D: get your mate to do the work, while you hold on to him

25.36

You have been asked to install a number of ceiling-mounted air-conditioning units in a large open-plan area, which has a good floor. It is best to do this:

- [ ] A: using stepladders
- [ ] B: using scaffold boards and floor stands
- [ ] C: using packing cases to stand on
- [ ] D: from a mobile tower

25.37

You have to drill through a wall panel that you suspect contains an asbestos material. What should you do?

- [ ] A: Ignore it and carry on
- [ ] B: Put on safety goggles
- [ ] C: Put on a dust mask
- [ ] D: Stop work and report it

25.38

When repairing an electrically-driven compressor, what is the minimum safe method of isolation?

- [ ] A: Pressing the stop button
- [ ] B: Pressing the emergency stop button
- [ ] C: Turning off the local isolator
- [ ] D: Locking off and tagging out the local isolator

25.39

When it is necessary to cut into an existing refrigerant pipe, should you:

- [ ] A: vent the gas in the pipework to atmosphere
- [ ] B: recover the refrigerant gas and make a record of it, then do the work
- [ ] C: work on the pipework with the refrigerant gas still in it
- [ ] D: not carry out the work at all, because of the risks

Answers: 25.35 = C, 25.36 = D, 25.37 = D, 25.38 = D, 25.39 = B

25.40

What is the FIRST thing that should be done when a new refrigeration system has been installed?

- [ ] A: It should be pressure and leak tested
- [ ] B: It should be filled with refrigerant
- [ ] C: It should be left open to air
- [ ] D: It should be turned off at the electrical switch

25.41

On water-cooled systems, the water in the system should be:

- [ ] A: replaced annually
- [ ] B: chemically treated
- [ ] C: properly filtered
- [ ] D: drinking water

25.42

What should you establish before entering a cold room?

- [ ] A: The size of the cold room
- [ ] B: The temperature of the cold room
- [ ] C: Whether the exit door is fitted with an internal handle
- [ ] D: The produce being stored

25.43

When a refrigerant leak is reported in a closed area, what should you do first before entering the area?

- [ ] A: Ventilate the area
- [ ] B: Establish that it is safe to enter
- [ ] C: Get a torch
- [ ] D: Wear safety footwear

25.44

When using a van to transport a refrigerant bottle, how should it be carried?

- [ ] A: In the back of the van
- [ ] B: In the passenger footwell of the van
- [ ] C: In a purpose-built container within the vehicle
- [ ] D: In the van with all the windows open

25.45

What safety devices should be fitted between the pipes and the gauges of oxy-propane welding equipment?

- [ ] A: Non-return valves
- [ ] B: On-off taps
- [ ] C: Flame retardant tape
- [ ] D: Flashback arresters

Answers: 25.40 = A, 25.41 = B, 25.42 = C, 25.43 = B, 25.44 = C, 25.45 = D

25.46

Which is the best and safest method of taking long lengths of copper pipe by van?

- [ ] A: Tying the pipes to the roof with copper wire
- [ ] B: Someone holding the pipes on the roof rack as you drive along
- [ ] C: Putting the pipes inside the van with the ends out of the passenger window
- [ ] D: Using a pipe rack fixed to the roof of the van

25.47

When using a blowtorch to joint copper tube and fittings in a property, a fire extinguisher should be:

- [ ] A: available in the immediate work area
- [ ] B: held over the joint while you are using the blowtorch
- [ ] C: used to cool the fitting
- [ ] D: available only if a property is occupied

25.48

When using a blowtorch, you should:

- [ ] A: stop using the blowtorch immediately before leaving the job
- [ ] B: stop using the blowtorch at least 1 hour before leaving the job
- [ ] C: stop using the blowtorch at least 2 hours before leaving the job
- [ ] D: stop using the blowtorch at least 4 hours before leaving the job

25.49

Why is it essential to take great care when handling oxygen cylinders?

- [ ] A: They contain highly flammable compressed gas
- [ ] B: They contain highly flammable liquid gas
- [ ] C: They are filled to extremely high pressures
- [ ] D: They contain poisonous gas

25.50

What part of the body could suffer long-term damage when hand-bending copper pipe using a spring?

- [ ] A: Elbows
- [ ] B: Shoulders
- [ ] C: Back
- [ ] D: Knees

Answers: 25.46 = D, 25.47 = A, 25.48 = B, 25.49 = C, 25.50 = D

HVACR: Services and Facilities

26.1

Where might you find information on the safe way to maintain the services, etc, in a building?

- A: The notice board
- B: The safety officer
- C: The local Health and Safety Executive office
- D: The Health and Safety File for the building

26.2

In the normal office environment, what should be the hot-water temperature at the tap furthest from the boiler, after running it for one minute?

- A: At least 15°C
- B: At least 35°C
- C: At least 50°C
- D: At least 100°C

26.3

What should be the maximum temperature for a cold-water supply, after running it for two minutes?

- A: 10°C
- B: 20°C
- C: 35°C
- D: 50°C

26.4

How is legionella transmitted?

- A: Breathing contaminated airborne water droplets
- B: Human contact
- C: Dirty clothes
- D: By rats' urine

26.5

When working in a riser, how should access be controlled?

- A: By a site security operative
- B: By those who are working in it
- C: By the main contractor
- D: By a Permit to Work system

26.6

What is the ideal temperature for legionella to breed?

- A: Below 20°C
- B: Between 20°C and 45°C
- C: Above 50°C
- D: Between 50°C and 100°C

Answers: 26.1 = D, 26.2 = C, 26.3 = B, 26.4 = A, 26.5 = D, 26.6 = B

26.7

What is required if there is a cooling tower on site?

- [ ] A: A formal logbook
- [ ] B: A Written Scheme of Examination
- [ ] C: Regular visits by the local authority Environmental Health Officer
- [ ] D: Inspections by the water supplier

26.8

Who should be informed if a legionella outbreak is suspected?

- [ ] A: The Health and Safety Executive
- [ ] B: The police
- [ ] C: A coroner
- [ ] D: The nearest hospital

26.9

Which of the following is the most likely place to find legionella?

- [ ] A: Drinking water
- [ ] B: Hot water taps above 50°C
- [ ] C: Infrequently used shower heads
- [ ] D: Local river

26.10

Which two of the following are pressure systems?

- [ ] A: Medium and high temperature hot water systems at or above 95°C
- [ ] B: Cold water systems
- [ ] C: Steam systems
- [ ] D: Office tea urns
- [ ] E: Domestic heating systems

26.11

When assembling a mobile access tower what major overhead hazard must you be aware of?

- [ ] A: Water pipes
- [ ] B: Cable trays
- [ ] C: False ceilings
- [ ] D: Suspended electric cables

26.12

You spill some oil on the floor and you do not have any absorbent material to clean the area. What should you do?

- [ ] A: Spread it about to lessen the depth
- [ ] B: Keep people out of the area and inform your supervisor
- [ ] C: Do nothing, it will eventually soak into the floor
- [ ] D: Warn other people as they tread through it

Answers: 26.7 = A, 26.8 = A, 26.9 = C, 26.10 = A,C, 26.11 = D, 26.12 = B

26.13

What should folding stepladders be used for?

- [ ] A: General access on site
- [ ] B: Short-term activities lasting minutes that don't involve stretching
- [ ] C: All work at height where a ladder cannot be used
- [ ] D: Getting on and off mobile towers

26.14

What is required before a pressure system can be operated?

- [ ] A: A Written Scheme of Examination
- [ ] B: A Permit to Work
- [ ] C: City & Guilds certification
- [ ] D: A minimum of two competent persons to operate the system

26.15

When should the use of a Permit to Work be considered?

- [ ] A: All high-risk work activities
- [ ] B: All equipment isolations
- [ ] C: At the beginning of each shift
- [ ] D: When there is enough time to complete the paperwork

26.16

Who should fit a padlock and tag to an electrical lock-out guard?

- [ ] A: Anyone working on the unit
- [ ] B: The engineer who fitted the lock-out device
- [ ] C: The senior engineer
- [ ] D: Anyone

26.17

You arrive on site and find the mains isolator for a component is switched off. What should you do?

- [ ] A: Switch it on and get on with your work
- [ ] B: Switch it on and check the safety circuits to see if there is a fault
- [ ] C: Contact the person in control of the premises
- [ ] D: Ask people around the building and if no one responds, switch it on and get on with your work

Answers: 26.13 = B, 26.14 = A, 26.15 = A, 26.16 = A, 26.17 = C

26.18

Before working on electrically powered equipment, what is the procedure to make sure that the supply is dead before work starts?

- [ ] A: Switch off and remove the fuses
- [ ] B: Switch off and cut through the supply with insulated pliers
- [ ] C: Test the circuit, switch off and isolate the supply at the mains board
- [ ] D: Switch off, isolate the supply at the main board, lock and tag out

26.19

When you are about to work on electrical equipment and the main isolator does not have a lock-out device, what is an alternative method of isolating the supply?

- [ ] A: Detach the mains power supply
- [ ] B: Place insulating material in the contactors
- [ ] C: There is no alternative to a lock-out device for isolating the supply
- [ ] D: Withdraw and retain the fuses and hang a warning sign

26.20

When arriving at an occupied building, who or what should you consult before starting work to find out about any asbestos in the premises?

- [ ] A: The building owner to view the Asbestos Register
- [ ] B: The building receptionist
- [ ] C: The building logbook
- [ ] D: The building caretaker

26.21

On cooling tower systems, the water in the system should be:

- [ ] A: replaced annually
- [ ] B: chemically treated
- [ ] C: chilled
- [ ] D: drinking water

26.22

When carrying out solvent welding on plastic pipework, what particular safety measure must you apply?

- [ ] A: The area must be well ventilated
- [ ] B: The supervisor must be present
- [ ] C: The area must be enclosed
- [ ] D: It must be done in daylight

Answers: 26.18 = D, 26.19 = D, 26.20 = A, 26.21 = B, 26.22 = A

26.23

What action should you take if a natural gas leak is reported in a closed area?

- [ ] A: Ventilate the area and phone the gas emergency service
- [ ] B: Establish whether or not it is safe to enter
- [ ] C: Turn off the light
- [ ] D: Wear safety footwear

26.24

Which of the following two actions should you take if a refrigerant leak is reported in a closed area?

- [ ] A: Ventilate the area and extinguish all naked flames
- [ ] B: Trace the leak and make a temporary repair
- [ ] C: Establish whether or not it is safe to enter
- [ ] D: Switch off the system
- [ ] E: No action as refrigerant gas is harmless

26.25

Which of the following should you not do when replacing the filters in an air-conditioning system?

- [ ] A: Put the old filters in a dustbin
- [ ] B: Follow a job-specific risk assessment and method statement
- [ ] C: Wear appropriate overalls
- [ ] D: Wear a respirator

26.26

Before starting work on a particular piece of equipment, who or what should you consult?

- [ ] A: The machine brochure
- [ ] B: The operation and maintenance manual for the equipment
- [ ] C: The manufacturer's data plate
- [ ] D: The storeman

26.27

After servicing a gas boiler, what checks must you make by law?

- [ ] A: For water leaks
- [ ] B: For flueing, ventilation, gas rate and safe functioning
- [ ] C: The pressure relief valve
- [ ] D: The thermostat setting

26.28

Before adding inhibitor to a heating system, what must you do?

- [ ] A: Check for leaks on the system
- [ ] B: Raise the system to working temperature
- [ ] C: Read the COSHH assessment for the product
- [ ] D: Bleed the heat emitters

Answers: 26.23 = A, 26.24 = A,C, 26.25 = A, 26.26 = B, 26.27 = B, 26.28 = C

26.29

How should Liquefied Petroleum Gas (LPG) cylinders be carried to and from premises in a van?

- [ ] A: In the back of the van
- [ ] B: In the passenger footwell of the van
- [ ] C: In a purpose-built container within the vehicle
- [ ] D: In the van with all the windows open

26.30

What should you do if, when carrying out a particular task, the correct tool is not available?

- [ ] A: Wait until you have the appropriate tool for the task
- [ ] B: Borrow a tool from the building caretaker
- [ ] C: Use the best tool available in the toolkit
- [ ] D: Modify one of the tools you have

26.31

What would you use to find out whether a wall into which you are about to drill contains an electric supply?

- [ ] A: A neon screwdriver
- [ ] B: A cable tracer
- [ ] C: A multimeter
- [ ] D: A hammer and chisel

26.32

Which type of power drill is most suitable for fixing a run of pipework outside in wet weather?

- [ ] A: Battery-powered drill
- [ ] B: Drill with 110 volt power supply
- [ ] C: Drill with 24 volt power supply
- [ ] D: Any mains voltage drill with a power breaker

26.33

Temporary continuity bonding is carried out before removing and replacing sections of metallic pipework to:

- [ ] A: provide a continuous earth for the pipework installation
- [ ] B: prevent any chance of blowing a fuse
- [ ] C: maintain the live supply to the electrical circuit
- [ ] D: prevent any chance of corrosion to the pipework

Answers: 26.29 = C, 26.30 = A, 26.31 = B, 26.32 = A, 26.33 = A

26.34

What is the procedure for ensuring that the electrical supply is dead before replacing an electric immersion heater?

- [ ] A: Switch off and disconnect the supply to the immersion heater
- [ ] B: Switch off and cut through the electric cable with insulated pliers
- [ ] C: Switch off and test the circuit
- [ ] D: Lock off the supply, isolate at the mains board, test the circuit and hang a warning sign

26.35

What is used to reduce 240 volts to 110 volts on site?

- [ ] A: Residual current device
- [ ] B: Transformer
- [ ] C: Circuit breaker
- [ ] D: Step-down generator

26.36

When positioning and erecting a stepladder, which of the following is essential for its safe use?

- [ ] A: It has a tool tray towards the top of the steps
- [ ] B: The restraint mechanism is spread to its full extent
- [ ] C: You will be able to reach the job by standing on the top step
- [ ] D: A competent person has positioned and erected the steps

26.37

What colour power outlet on a portable generator would supply 240 volts?

- [ ] A: Black
- [ ] B: Blue
- [ ] C: Red
- [ ] D: Yellow

26.38

When removing some panelling, you see a section of cabling with the wires showing. What should you do?

- [ ] A: Carry on with your work, trying your best to avoid the cables
- [ ] B: Touch the cables to see if they are live, and if so refuse to carry out the work
- [ ] C: Wrap the defective cable with approved electrical insulation tape
- [ ] D: Only work when the cable has been isolated or repaired by a competent person

26.39

When must a shaft or pit be securely covered or have double quard-rails and toe-boards installed?

- [ ] A: Fall height of 1.0 metre
- [ ] B: When there is a risk of injury from someone falling
- [ ] C: Fall height of 2.5 metres
- [ ] D: Fall height of 3.0 metres

Answers: 26.34 = D, 26.35 = B, 26.36 = B, 26.37 = B, 26.38 = D, 26.39 = B

26.40

What is the correct action to take if natural gas is detected in an underground service duct?

- [ ] A: No action, if it is not harmful
- [ ] B: Evacuate the duct
- [ ] C: Carry on working but do not use electrical equipment
- [ ] D: Carry on working until the end of the shift

26.41

What should you think of first when planning to work in a confined space?

- [ ] A: Whether the job has been priced properly
- [ ] B: Whether sufficient manpower has been allocated
- [ ] C: Whether the correct tools have been arranged
- [ ] D: Whether the work can be done without entry into the confined space

26.42

Which two of the following should a person who is going to work alone carry out to ensure their safety?

- [ ] A: Register their presence with the site representative before starting work
- [ ] B: Ensure their timesheet is accurate and countersigned
- [ ] C: Establish suitable arrangements to ensure the monitoring of their well-being
- [ ] D: Notify the site manager of the details of the work
- [ ] E: Only work outside of normal working hours

26.43

How should an adequate supply of breathable fresh air be provided in a confined space in which breathing equipment is not being worn?

- [ ] A: An opening in the top of the confined space
- [ ] B: Forced mechanical ventilation
- [ ] C: Natural ventilation
- [ ] D: An opening at the bottom of the confined space

Answers: 26.40 = B, 26.41 = D, 26.42 = A,C, 26.43 = B

26.44

What are the two main safety considerations when using oxyacetylene equipment in a confined space?

- [ ] A: The hoses may not be long enough
- [ ] B: Unburnt oxygen causing an oxygen-enriched atmosphere
- [ ] C: The burner will be hard to light
- [ ] D: Wearing the correct goggles
- [ ] E: Flammable gas leak

26.45

What precaution should be taken to protect against lighting failure in a confined space?

- [ ] A: Remember where you got in
- [ ] B: Ensure it is daylight when you do the work
- [ ] C: Each operative should carry a torch
- [ ] D: Secure a rope near the entrance and trail it behind you so that you can trace your way back

26.46

To prevent unauthorised access to an unoccupied plant or switchgear room, what must you ensure?

- [ ] A: That the access door is locked
- [ ] B: That a sign is posted
- [ ] C: That the power supply is isolated
- [ ] D: That a person is posted to prevent access

26.47

While working on your own and tracing pipework in a building, the pipes enter a service duct. You should:

- [ ] A: go into the service duct and continue to trace the pipework
- [ ] B: ask someone in the building to act as your second person
- [ ] C: put on your Personal Protective Equipment and carry on with the job
- [ ] D: stop work until a risk assessment has been carried out

Answers: 26.44 = B,E, 26.45 = C, 26.46 = A, 26.47 = D

26.48

What is the first thing you should do after getting on to the platform of a correctly erected mobile tower?

☐ A: Check that the brakes are locked on

☐ B: Check the mobile tower to make sure that it has been correctly assembled

☐ C: Close the access hatch to prevent falls of personnel, tools or equipment

☐ D: Make sure that the tower does not rock or wobble

26.49

Before using a ladder at work, you notice that the maker's label says that it is a 'Class 3' ladder. What does this mean?

☐ A: It is for domestic use only and must not be used in connection with work

☐ B: It is of industrial quality and can be used safely

☐ C: It has been made to a European Standard

☐ D: It is made of insulating material and can be used near to overhead cables

26.50

When accessing a roof by ladder, the only available point of rest for the ladder is a run of plastic gutter. What should you do?

☐ A: Ensure the point of rest is over one of the gutter support brackets

☐ B: Rest the ladder against the run of gutter, climb it and quickly tie it off

☐ C: Use a proprietary 'stand-off' device that allows the ladder to rest against the wall

☐ D: Position the ladder at a shallow angle so that it rests below the gutter

Answers: 26.48 = C, 26.49 = A, 26.50 = C

Northern Ireland legislation

Legislation in Northern Ireland differs from that in the rest of the UK. The following questions highlight differences in legislation which apply to candidates in Northern Ireland. Please be aware that these differences will not be shown in the test. All candidates will be tested on legislation relating to the rest of the UK only.

Northern Ireland legislation

Section 1 – General Responsibilities

1.5

The Health and Safety at Work Order (NI) 1978 places legal duties on:

- ☐ A: employers only
- ☐ B: operatives only
- ☐ C: all people at work
- ☐ D: self-employed people only

1.8

Why is the Health and Safety at Work Order (NI) 1978 important to you? Give two answers.

- ☐ A: It tells you which parts of the site are dangerous
- ☐ B: It must be learned before starting work
- ☐ C: It requires your employer to provide a safe place to work
- ☐ D: It tells you how to do your job
- ☐ E: It puts legal duties on you as an employee

1.12

As an employee, which of these is not your duty under the Health and Safety at Work Act?

- ☐ A: To look after your own health and safety
- ☐ B: To look after the health and safety of anyone else who might be affected by your work
- ☐ C: To write your own risk assessments
- ☐ D: Not to interfere with anything provided for health and safety

Answers: 1.5 = C, 1.8 = C, E, 1.12 = C

Section 16 – Supervisory and Management

16.3

Which form must be displayed on projects where the Construction (Design and Management) Regulations (NI) apply?

- [ ] A: Form NI9
- [ ] B: Form NI10 (Rev)
- [ ] C: Form 11
- [ ] D: Form 12

16.4

Under the Construction (Design and Management) Regulation (NI)s, who is responsible for ensuring notification to the Health and Safety Executive of the project?

- [ ] A: Client
- [ ] B: Designer
- [ ] C: Planning supervisor
- [ ] D: Principal contractor

16.10

Following a reportable dangerous occurrence, when must the Health and Safety Executive be informed?

- [ ] A: Within 1 day
- [ ] B: Within 5 days
- [ ] C: Within 10 days
- [ ] D: Immediately

16.37

The Construction (Health, Safety and Welfare) Regulations (NI) 1996 require a supported excavation to be inspected:

- [ ] A: every 7 days
- [ ] B: at the start of every shift
- [ ] C: once a month
- [ ] D: when it is more than 2 metres deep

16.48

The Management of Health and Safety at Work Regulations (NI) 2000 require risk assessments to be made:

- [ ] A: for all work activities
- [ ] B: when there is a danger of someone getting hurt
- [ ] C: when more than five people are employed
- [ ] D: where an accident has happened previously

16.54

The Work at Height Regulations (NI) require inspections of scaffolding to be carried out by:

- [ ] A: the site manager
- [ ] B: the scaffolder
- [ ] C: the safety adviser
- [ ] D: a competent person

Answers: 16.3 = B, 16.4 = C, 16.10 = D, 16.37 = B, 16.48 = A, 16.54 = D

Section 16 – Supervisory and Management continued

16.55

Under the requirements of the Construction (Design and Management) Regulations (NI), what has to be displayed on a construction site?

- [ ] A: Notice of application to erect hoardings
- [ ] B: Notice of the Health and Safety Commission's address
- [ ] C: Form NI10 (Rev)
- [ ] D: A statement by the client

16.67

The Health and Safety at Work Order and any regulations made under that Order are:

- [ ] A: advisory to companies and individuals
- [ ] B: legally binding
- [ ] C: good practical advice for the employer to follow
- [ ] D: not compulsory, but should be complied with if convenient

Section 17 - Demolition

17.15

Where demolition is to be carried out close to overhead cables, who should be consulted?

- [ ] A: The Health and Safety Executive
- [ ] B: NI Fire Service
- [ ] C: The local electricity supply company
- [ ] D: The National Grid

17.34

Plant and equipment needs to be inspected and the details recorded by operators:

- [ ] A: daily at the beginning of each shift
- [ ] B: weekly
- [ ] C: monthly
- [ ] D: every 3 months

Answers: 16.55 = C, 16.67 = B, 17.15 = C, 17.34 = A

Acknowledgements

CITB-ConstructionSkills wishes to acknowledge the assistance offered by the following organisations in the development of health and safety testing:

Health and Safety Test Question Sub-Committee

CITB (NI)

Construction Employers Federation Limited (CEF NI)

Driving Standards Agency

Heating and Ventilating Contractors' Association (HVCA)

Highways Agency

Joint Industry Board – Plumbing, Mechanical and Electrical Services

Lift and Escalator Industry Association (LEIA)

Management Board of the Construction Skills Certification Scheme

National Demolition Training Group

Scottish and Northern Ireland Joint Industry Board for the Plumbing Industry

Notes

Notes

Notes

Notes

Notes